The Natural Savage
Discovering the Human Animal

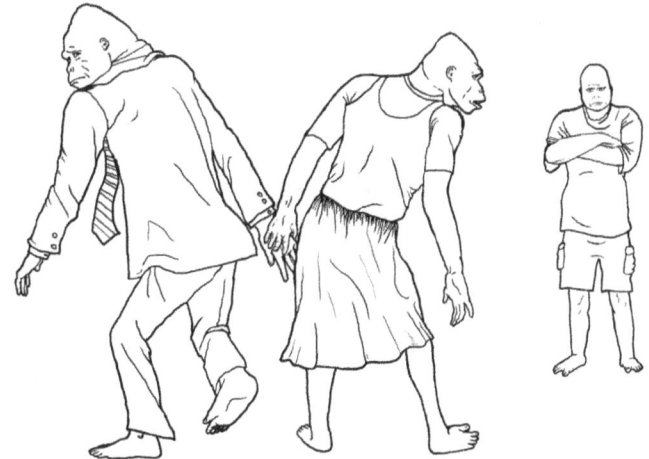

Anthony Hernandez

The Natural Savage
Discovering the Human Animal

Anthony Hernandez

All rights reserved.
Copyright © 2012

Illustrations by Boni Meredith (boni_m@yahoo.com)

Hernandez, Anthony 1968-
The Natural Savage/Anthony Hernandez
    ISBN: 978-0-9855793-5-7 (paperback)
    ISBN: 978-0-9855793-6-4 (Amazon® Kindle®)

Advance Praise for *The Natural Savage*

"Anyone who wants to know where we come from and how ancient instincts are guiding everything we do must read *The Natural Savage*. Anthony Hernandez makes human evolution easy to understand and peels back the layers of civilization to reveal the ultimate simplicity of life and how and why we do the things we do. Knowing how our past shapes the present is the key to moving toward a future of ease, abundance, and peace."

Caterina Rando, MA, MCC
Keynote speaker, success coach
Author, *Learn to Power Think*
www.caterinar.com

Anthony Hernandez's book, *The Natural Savage*, is a self-help book for those who are curious and who are interested in the "bigger picture." More than providing pat answers, this book is an invitation to reflect, to be inquisitive, and to ask questions. The book is rich in the breadth and scope of the literature reviewed and engages the reader by contemplating the complexity and simplicity of many facets of human behavior from the vantage point of what scientists are learning about our animal cousins and wondering about what makes us tick.

John Bilorusky, PhD
President and Faculty Member,
Western Institute for Social Research
www.wisr.edu

The Natural Savage
Discovering the Human Animal

Table of Contents

Advance Praise	i
Dedication	xv
Acknowledgements	xvii
Introduction	xix

PART ONE
Foundations

1: Assumptions — 3
- The Scientific Muddle — 3
- Evolution — 4
- To Serve Man — 7
- Creation — 9

2: Reducto ad Absurdum — 11
- Life at a Glance — 11
- The Six Functions of Life — 12
  - Predator Avoidance — 12
  - Group Status — 13
  - Sustenance — 15
  - Shelter — 16
  - Reproduction — 16
  - Death — 17

3: The Other Chimpanzees — 21
- Tearing Down the Wall — 21
- Common Heritage — 22
  - The Great Divide — 23
  - From Ape to Man — 24
- Anatomy — 24
  - Limbs — 24
  - Skulls — 25
  - Vision — 25
  - Consciousness — 26

	Diet	26
	Chimpanzee	27
	Bonobo	27
	Human	27
	Range	28
	Society	28
	Chimpanzee	28
	Bonobo	29
	Human	30
	Tools	30
	Language	31
	Reproduction	31
	Chimpanzees	31
	Bonobos	32
	Conclusion	33
4:	**All God's Creatures**	**35**
	The Stuff of Stars	36
	Ex Unus Plures	36
	Intelligence	38
	The Triune Brain	38
	The Frontal Lobes	39
	Is Bigger Better?	40
	Measuring Intelligence	40
	Perception	41
	Emotions	43
	Happiness	44
	Sadness	44
	Aggression	44
	Fear	45
	Love	45
	Grief and Despair	46
	Guilt and Shame	46
	Society	46
	Language	47
	Child Rearing	47
	Altruism	48
	Play	49
	Civilization?	49
	Conclusion	50

5: What is Reality? — 51
- Perception at a Distance — 51
- A Living Computer — 52
- The Process of Realization — 53
  - Early Emotions — 53
  - Thoughts — 54
  - Emotions — 55
  - Actions — 55
  - Results — 56
- Reality Defined — 56

PART TWO
Predator Avoidance

6: Rise of the Prey — 59
- Our Evolutionary Tree — 59
- Humans as Prey — 61
  - Monkey Predators — 62
  - Prometheus Rising — 62
  - Primeval Evidence — 63
- Predator/Prey Behavior — 63
- Fear in Action — 64
  - Fight or Flight? — 65
- The Bottom Line — 66

7: Predation 101 — 67
- The Predation Process — 68
  - Searching — 69
  - Detection — 70
  - Acquisition — 71
  - Firing Solution — 72
  - Attack — 73
  - Kill — 74
- Coping — 74
  - Asymmetry — 75
  - Built-In Fears — 75
  - Reflexes — 75
  - Causality — 76
  - Confirmation Bias — 76
  - Predicting the Future — 76

	Universal Distress Noises	77
	Good Night	77
	What Me Worry?	78
8:	**Programming the Human Animal**	**79**
	Nature or Nurture?	79
	The Bell Curve	80
	Canine Connections	82
	Rhythm	83
	Aww!	83
	Big Head Narrow Pelvis	84
	Learning 101	86
	Methods	88
	Motion	88
	Language	89
	Emotions	89
	Cheater Detection	90
	Security vs. Independence	91
	Timing	92
	Honesty	92
	Consistency	92
	Education	93
	Play	94
	Conflict	94
	The Programmed Animal	95
9:	**Communication**	**97**
	Why Sound?	98
	Why Language?	98
	The Anatomy of Language	99
	Physical Anatomy	99
	Tonal Anatomy	100
	Beyond Verbal Communication	100
	Language Evolution	101
	Learning to Speak	106
	Language Genetics	107
	Literacy	108

PART THREE
Status

10: The High Price of Fitting In — 113
- Civilization — 116
- Selection Pressures — 116
  - From Trees to Towers — 117
- Getting Along — 119
  - Friendship First — 120
  - Under Pressure — 121
  - Reciprocity — 122
- Mental Illness — 124

11: It's Good to be King — 127
- Inherent Inequality — 127
  - Haves & Have-Nots — 128
  - Castes — 129
  - Dynasties — 130
- Alpha — 131
- Omega — 132
  - Submissive Behaviors — 134
- Jockeying for Position — 136
  - Stable vs. Unstable Hierarchies — 136
  - Politics — 137
  - Ladder Climbing — 139
- Roaring Mice — 140
  - Friends in Low Places — 140
  - Ritualized Combat — 140
- Religion — 141
- King for a Day — 142
  - Tiny Fiefdoms — 142
  - Faking It — 143

12: War & Conquest — 145
- Us vs. Them — 146
- Violent Nature — 147
- Animals at War — 148
- Humans at War — 149
  - Why Fight? — 151
  - A Very Brief History of Conquest — 153

	Genocide	155
	Armed & Dangerous	156

13: Status at Work — 159
- The Profession Deception — 160
- The Universal Resource — 161
- Tribes within Tribes — 162
- Women at Work — 166

14: Status at Home — 169
- Sexual Division — 169
  - Why Women? — 170
  - Win-Win — 171
  - Monogamy — 171
- Uneven Investments — 172
  - Investment vs. Power — 173
  - Fundamental Errors? — 173
- Conflict at Home — 175
  - In the Womb — 175
  - Weaning and Walking — 176
  - Sibling Rivalry — 176
  - Parents vs. Children — 176
  - Spousal Conflict — 177
- Friends — 180

PART FOUR
Three Hots and a Cot

15: The Omnivore's Dilemma — 183
- Complex Larders — 184
- Picky Eaters — 185
  - Vive la Cuisine — 186
  - Confronting the Unknown — 187
  - Finicky Children — 188
  - When Things Go Wrong — 189
  - The Dietary Conundrum — 190
- Gourmets vs. Gourmands — 190
- Other Eating — 191
- Vegetarianism? — 192
- Entitlement — 193

Contents

16:	**What's for Dinner?**	**195**
	Agriculture	195
	Roots	196
	Property	196
	Specialization	197
	Restriction	197
	Inequality	198
	Food Poisoning	200
	Changing Diets	201
	Fat	202
	Sugar	203
	Salt	204
	Ancient Palates	204
	At All Costs	205
	Reason for Hope	205
17:	**Bringing Home the Bacon**	**207**
	Bagging Dinner	207
	On the Hunt	210
	The Gambler's Fallacy	210
	Confirmation Bias	211
	Seek and Ye Shall Find	212
	Quantity vs. Quality	213
18:	**The Natural Savage at Home**	**215**
	Home Sweet Home	215
	Designated Spaces	217
	Cavemen?	218
	Modern Savannas	218
	Good Fences	219
	Teeming Multitudes	219
	SIDS	220
	Sharing Our Homes	220
	Morality	221
	Cooperation Displays	222
	Culture	223

PART FIVE
Fruitful Multiplication

19:	**Of Birds and Bees**	**227**
	No Looking Back	228
	Why Sex?	228
	Natural Selection	229
	Reproductive Strategies	231
	No Contact	231
	Single Queen	231
	Polygyny	232
	Serial Monogamy	232
	Monogamy	233
	Polyandry	233
	Cuckoldry	234
	Incest	234
	Internal vs. External Fertilization	235
	Mixed Strategies	235
	Sexual Cannibalism	236
	Productivity	237
	Vasopressin	238
	Rolling the Dice	238
20:	**Sexy Monkeys**	**239**
	Fidelity	240
	Attraction	242
	Physical	242
	Biochemical	244
	Demographics/Psychographics	246
	Handicaps	248
	Economics	249
	Lust & Sex	251
	Concealed Ovulation	251
	Male Equipment	252
	Female Equipment	252
	Orgasm	253
	Love	253
	Never Enough	253

Contents | xi

21:	**Courtship 101**	**255**
	Initial Attraction	256
	Attraction Signals	259
	Consummation	261
	Pre-Copulation	262
	Copulation	263
	Post-Copulation	264
	Keeping the Faith	264
	Games People Play	267
	In Summary	270
22:	**Neither Straight nor Narrow**	**271**
	Sex on the Brain	272
	Sex in Practice	273
	Frequency	273
	Orgasm	274
	Starting Young	274
	Contraception	276
	Serial Monogamy	277
	Beyond Vanilla	278
	Fantasy	279
	Masturbation	279
	Pornography	279
	Oral Sex	280
	Anal Sex	280
	BDSM	280
	Homosexuality	280
	Prostitution	282
	Cheating	283
	STDs	284
23:	**Morality vs. Nature**	**285**
	Monogamy	286
	Adultery & Divorce	287
	Sex & Violence	287
	Religion	288
	The Bible	288
	The Koran	291
	Mormonism	292
	Buddhism	293

Hinduism		293
Secular Law		294
Lessons Learned		294
Homosexuality		295

PART SIX
The Great Unknowns

24:	**Death: The Final Frontier?**	**299**
	Life Beyond Reproduction	300
	Repair or Recycle?	301
	Universal Deterioration	302
	Life without Death	303
	Design or Accident?	305
	Self-Awareness	306
	No Waste	306
	The Ascension Fallacy	307
	How Little We Know	310
	Afterlife?	311
	Eternal Life	311
	Life after Death	312
25:	**Putting it all Together**	**313**
	Animal Urges	313
	The Substance of Style	315
	Choose Joy	315
	Live on Purpose	316
	Acknowledge Others Often	316
	Ask for What You Want	316
	Be Willing to be Uncomfortable	316
	Explore New Possibilities	317
	Maintain a Positive Disposition	317
	Take Small Actions	317
	Openly Express Your Gratitude	317
	Have Some Fun	318
	Smile, Laugh, and Love More	318
	Look for the Ease	318
	Always Expect Success	318
	You Have the Power	318
	The Meaning of Humanity	319

| *26:* | **Who Knows?** | **321** |

Resources **323**
    Books 323
    Web Sites 325

xiv | *The Natural Savage*
Discovering the Human Animal

Dedication

This book is dedicated to my son Logan, who never ceases to amaze me with his humor, brilliance, and love.

xvi | *The Natural Savage*
Discovering the Human Animal

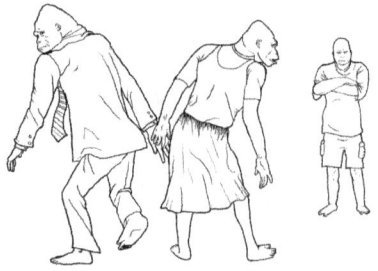

Acknowledgements

No project of this magnitude can occur in a vacuum and *The Natural Savage* is certainly no exception. I need to thank the many people who loved and supported me from the very beginning.

I owe my largest debt of gratitude to my family who endured night after night of me sitting in the next room lost in my own little world as I typed these pages. This book would simply not have happened without their unconditional love and support. It is no easy thing being so near yet so far and I can only guess that it was that much harder on them. Thank you.

Jared Diamond's book *The Third Chimpanzee* is easily one of the most amazing books I've ever read. That book inspired me to continue the research that went into this book and gave me many ideas on what direction that research should take.

Many, many thanks to John Bilorusky at the Western Institute for Social Research. John reviewed my initial outline and reading list, recommended additional reading, and kept firing tough questions at me. His involvement with this project was invaluable.

My friend Andrew Davis and I have enjoyed many talks that helped form the germ for this book and others coming in the *Savage* series.

Larry Loebig is one of my mentors. His guidance and support have kept me going through the tough spots and his constant pushing me to excel continue to yield positive results.

Jay Conrad Levinson is the father of the worldwide Guerrilla Marketing revolution. He and his wife Jeannie and daughter Amy made the introductions that snowballed into my being able to both write the *Savage* series of books and get them published. I can't thank the Levinsons enough.

My partner Jennifer Cummings and I enjoy many discussions about chimpanzees, bonobos, humans, and the similarities and differences between these species. OK, so we've focused mainly on the similarities, but still...

Thank you all.

xviii | *The Natural Savage*
Discovering the Human Animal

Introduction

Welcome and thank you for choosing to read *The Natural Savage*. This book looks at some of the many similarities between humans and animals, and includes a few comparisons and contrasts with traditional psychology to deliver my ideas about what—and who—we humans are by nature.

To fully understand *The Natural Savage*, you need to know a little about both why I wrote it and why I wrote it the way I did. The best way for me to answer these questions is to share portions of my life's story, which will expose both my motivations and the reasons behind those motivations. It may also shed some light on the many biases that I, like any author, put into this book. Knowing where the author is coming from is crucial to getting the most from a book, and *The Natural Savage* is no exception.

As a child, I paged through Desmond Morris's *Manwatching* many times and devoured *National Geographic* articles. I particularly remember reading about the Leakeys discovering proto-human footprints in the Serengeti and the Lucy skeleton. Jane Goodall's work with chimpanzees also fascinated me because of her accounts of chimpanzee families and familial/tribal interactions that struck me as being extremely human-like in nature. I also remember being enthralled by nature shows that, to me, showcased both the similarities between humans and animals and the very different behaviors of predators and prey animals.

Those early interests embarked me on a lifelong quest to find the answers (or at least my answers) to life's biggest questions: who we are, why we're here, and why people do some of the things we do. My attempts to answer the latter question began during a difficult childhood that prompted me to ask, "If people can harm those closest to them, what are they capable of against total strangers?"

As a young adult, I responded to the challenges I faced while growing up in two ways: First, I decided that the best defense could possibly be a good offense. Second, I withdrew from the world as much as possible and prided myself on my cold and aloof reliance on logic over emotion. *Star Trek's* Mr. Spock was both my hero and someone I attempted to emulate.

XX | The Natural Savage
Discovering the Human Animal

My approach to the world looked good in theory, but life has a way of tinkering with even the best-laid plans. Despite my self-imposed isolation, I met a girl who once forced me to sit still and watch a sunset—an experience I credit with opening my eyes to a whole new side of life. Around that time, I also joined the US Coast Guard Reserve where I got my fill of military life while also serving society in various constructive capacities.

A seemingly minor event in high school proved pivotal in my early twenties: My probability and statistics teacher tossed a coin during a lesson. He then asked whether the result of that toss would affect the outcome of a subsequent toss (obviously not). That lesson saved my life. On two occasions, I was prepared, gun in mouth, to commit suicide because I felt unable to face my pent-up emotions and deepening depression. Both times, I remembered the coin toss example, and refrained from pulling the trigger. After all, killing myself would only deprive me of the chance to flip another of life's coins.

With suicide out of the question, I pondered high-profile crimes such as Dan White's infamous "Twinkie defense" where people use their pasts to justify or mitigate horrible later acts. I realized two things: First, my past probably qualified me to literally get away with murder. Second, I had the choice to kill myself and/or others—or not. Right there, I learned that past performance does not dictate the future, as any mutual fund commercial will remind us, because I have the power of choice, which in turn has consequences.

Violence could not solve my problems. So what could? This is where memories of watching a sunset came to my rescue. Maybe—just maybe—there is a whole positive nurturing aspect of life that I'd been missing out on. But how to go about learning about—and implementing lessons from—a side of life that I knew next to nothing about?

I sought out self-help books, counseling, and therapy. Each source had good things to say about honoring my inner child, loving myself, and that my experiences and feelings were hardly unique. This helped explain my feelings and reassured me that I was neither crazy nor to blame for my childhood. Still, everyone I turned to at that time spoke in terms of fixing and healing. The analogy that comes to mind is dropping a glass. One can't blame the glass for shattering; however, the glass remains broken and in need of repair or discarding. The message seemed obvious: One or more things were wrong with me and needed fixing.

Around this time (late 80s/early 90s), I began hearing about budget cuts in mental health programs. I marveled at the vast numbers of people diagnosed with various mental problems and just had to wonder: Is the human race really this sick?

Self-help aside, I was also pursuing my interests in quantum physics and cosmology. I learned about concepts such as relativity (the answer to a question depends on your point of view), the Heisenberg Uncertainty Principle (a phenomenon is only measurable to a certain point), Shrödinger's Cat (events without observed outcomes exist in all possible states at once), and the observer effect (the observer is part of the equation and therefore part of the answer) that continue to intrigue me.

Enter books such as *Guerrilla Marketing* (which I studied because I've been self-employed most of my adult life) and motivational programs, such as Caterina Rando's *Success With Ease*, whose message was just as clear as my earlier sources: Think in a certain way, do certain things, and the desired outcome will occur. The techniques used to attain the desired mindset ranged from "just do it" to tools like affirmations and meditation. Follow these rules, the programs promised, and you will get these results. And they worked, to a point: My income increased, and I have yet to go hungry from lack of effective marketing despite any inherent laziness.

Human evolution, predator vs. prey behavior, group dynamics and hierarchies, my gut feeling that humans can't be as mentally ill as the numbers suggest, rejection of what little I knew of classical psychology, quantum physics, self-help/motivation... everything started to click. These subjects plus the concept of core beliefs gleaned from various sources finally had me asking, "What if everything good or bad that happens to us occurs not because something is wrong with us but because we're perfectly OK?"

For example, one friend of mine called her father every time something good happened in her music career—the same man who had punished her by repeatedly putting her musical instruments on the street with "Free!" signs. Another woman I spoke with wanted to marry a nice man and become a housewife and mother. Her profession? Prostitute.

The little psychology I knew from my own limited experiences in therapy would approach these women from a "What's wrong and how do we fix them?" standpoint. Indeed, the very word *therapy* means, "the treatment of disease or disorders, as by some remedial, rehabilitating, or curative process; a curative power or quality." The term *psychotherapy* means, "the treatment of psychological disorders or maladjustments by a professional technique, as psychoanalysis, group therapy, or behavioral therapy." (definitions obtained from www.dictionary.com).

The National Institute for Mental Health estimates that over 25% of the American population suffers "from a diagnosable mental disorder in a given year." The Agency for Healthcare Research and Quality reports that 32.6 million Americans (about 11%) purchased at least one prescribed psychotherapeutic drug in 2004. Forty-five million people are lifetime users of both non-prescribed pyschotherapeutic drugs and alcohol, according to the Substance Abuse and Mental Health Services Administration. The above figures do not include addiction to nicotine, or illegal psychoactive drugs, such as cocaine or methamphetamine.

These astonishing numbers seem to indicate that humans truly are an extremely sick species. If this is true, then I must wonder how we evolved this far, when individuals in other species are swiftly eliminated upon showing any weakness or deficiency. One alternative explanation may be that viewing people as "mentally ill" is fundamentally wrong, as commonly applied.

It finally dawned on me that both human behavior and many of our perceived mental disorders (except those caused by physical injury or illness) could be explained by comparing the human animal to our fellow animals. Going one step further by seeing humans as prey animals instead of predators gave me the building blocks for the model I first wrote about in *The Enlightened Savage*. These explorations also kindled an intense interest in both evolutionary psychology and sociobiology.

Both fields explain the "why," but seem to leave open the question of how to use that knowledge for practical purposes.

The motivational and self-help programs I studied did a great job of presenting models of behavior that could lead to results, while lacking scientifically solid explanations of the root causes for these behaviors. Think of a car owner's manual: It describes how to use the car's controls to get the desired results with no more than an occasional cursory explanation of how the car actually works.

Seeing humans as prey animals from the middle of the food chain at best provides both the "how" and "why." The human brain's primary job is to be right, because being wrong increases the risk of dying a premature death. Dying prematurely goes against the primary reasons for living, those being to live as long as possible and propagate the species.

Life for prey animals revolves around one simple mission: not getting killed and eaten. They must learn how to avoid their predators, and then keep following the same pattern over and over again. The two women I mentioned above were taught that they did not deserve what they wanted. They subconsciously interpreted those lessons as survival instructions: Behave this way or risk being killed and eaten. They then responded to those lessons by doing everything in their power to keep themselves from realizing their dreams. According to the model I created and presented in *The Enlightened Savage*, both women are perfectly healthy, because they acting in accordance with their deepest mental programming.

I use this model in my coaching. To date, it explains any behavior I've examined where I knew the person's history. In fact, I have yet to see a single example that either disproves my model or forces significant changes. In fairness, this either means that my model is rock solid, or that my own biases are too strong to allow me to see the disproof right under my nose. I choose to believe the former, at least for now.

Having built my model and written *The Enlightened Savage* to both present that model and some ideas for putting it to positive pragmatic use, my next obvious question was "What now?" *The Enlightened Savage* is a great book built on solid science (if I dare say so myself), but is not a scientific book at heart.

The answer is clear: I am continuing my research into the intersection between humans and other animals, and exploring my model of human nature and behavior in depth. I am exploring behind the scenes of the summary contained in *The Enlightened Savage*, and will present my findings and opinions in this and future books. My hope is that people who read *The Enlightened Savage* will be inspired to pick up future *Savage* books to continue their own explorations. The fact that you're reading this book proves that my hope was not in vain. Since it ponders the question of, "What are humans really like?" it seemed only natural to give it the title of *The Natural Savage*.

My lifelong search for the answers life's biggest questions has completely transformed me. I no longer ask myself, "What's wrong with me?" because I finally know that I'm fine. All of the turmoil I created in my life happened because I am perfectly healthy. I've been working on my self esteem for decades, and the last few years have revolutionized how I see myself. It's amazing to look at pictures of myself at 20 years of age and see a paragon of physical fitness while realizing how rotten I felt on the inside. Today, my body may not look so good, but my mind is the best it's ever been.

If nothing else, writing these books is motivating me to continue learning and growing. If reading them is doing the same for you, dear reader, then I have truly accomplished my mission.

Future titles in this series will include:

- *The Social Savage*, a detailed exploration of human relationships with a focus on hierarchies.

- *The Romantic Savage*, which will look at a very special subset of the human relationship spectrum (coming 2013).

- *The Divine Savage*, my attempt to answer no less a question than "Is there a God?" and its corollary "Do humans possess souls, or are we nothing but biological creatures?" Order your copy today at www.dawnstarbooks.com.

I hope you both enjoy and learn from this book.

Anthony Hernandez

One very important final note: As of this writing, I am neither a doctor, nor a mental health professional. I joke that I can't even diagnose a bee sting if I watch it happen, but that's the literal truth. Nothing in my model or in any of the *Savage* books is intended as any sort of medical advice or diagnosis whatsoever. I strongly recommend that you consult a doctor or licensed mental health professional if you feel the need.

xxvi | *The Natural Savage*
Discovering the Human Animal

PART ONE

Foundations

2 | *The Natural Savage*
Discovering the Human Animal

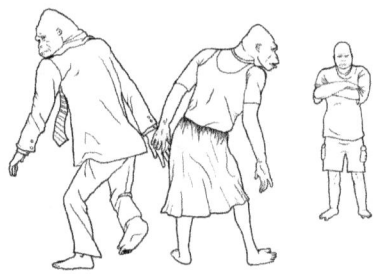

Chapter 1

Assumptions

> *That life is worth living is the most necessary of assumptions, and, were it not assumed, the most impossible of conclusions.*
> George Santayana

It is only fair that I begin this book by describing the core assumptions and biases that lie beneath the many assertions I make herein. Notice my use of the term *assertions* instead of *facts*. You, dear reader, need not accept anything in these pages as truth in order to benefit from this book. All I ask is that you read with an open mind and keep asking yourself one of the most profound questions ever asked: What if?

The Scientific Muddle

> *The world always makes the assumption that the exposure of an error is identical with the discovery of truth—that the error and truth are simply opposite. They are nothing of the sort. What the world turns to, when it is cured of one error, is usually simply another error, and maybe one worse than the first one.*
> H.L. Mencken

My dictionary defines a *fact* as something that actually exists; reality; truth. To paraphrase Mike Fortun and Herbert J. Bernstein's *Muddling Through: Pursuing Science and Truths in the Twenty-First Century*, it is tempting to believe that the scientific method rests on the foundation of hard, immutable facts that learned people collect and assemble into new theories or conclusions that eventually become facts in their own right. Unfortunately, things aren't quite that simple. Science is rife with disagreement and controversy, and one can often interpret the same set of data in several ways.

Consider that several famous scientists reached completely different conclusions about how our solar system works after observing the same nine planets moving across the same sky in the same way. Today, we accept the idea of a *heliocentric* (Sun-centered) solar system, whose planets move in simple elliptical orbits as truth, in the same way that our ancestors accepted the idea of a *terracentric* (Earth-centered) solar system, whose planets moved in layered circular orbits called epicycles as truth. Humanity's definition of truth is constantly being updated as science advances its own models. There are, of course, notable exceptions such as the Flat Earth Society, whose members contend that our planet is indeed flat with all evidence to the contrary being part of some great cover-up.

Things get even more interesting when science challenges the conventional wisdom, morals, or religious teachings of its day. Galileo Galilei, Nicolaus Copernicus, and Sir Isaac Newton were all accused of heresy. Charles Darwin's *On The Origin of Species* was received with great controversy. The random evolution versus intelligent design debate is alive and well today.

My first core assumption is that science, like the human species, is constantly evolving, and that new discoveries are forcing us to reconsider or even reject prior models. I therefore ask you to read this book with an open mind and accept my ideas, not as truth, but as the (hopefully) interesting and thought-provoking model they form.

> *Facts are not found, but made. The scientific method does not discover truth, it produces it.*
> Mike Fortun & Herbert J. Bernstein

> *I have learned throughout my life as a composer chiefly through my mistakes and pursuits of false assumptions, not by my exposure to founts of wisdom and knowledge.*
> Igor Stravinsky

Evolution

About 13.7 billion years ago (that's 13,700,000,000), an infinitesimal cosmic singularity exploded. Where this singularity came from and why it exploded remain mysteries, but the by-products of that event coalesced into the matter and energy that comprise the entire Universe. Look around you: Everything you see, hear, feel, smell, and taste is quite literally the stuff of stars.

Life on Earth began about 3.5 billion (3,500,000,000) years ago. Single-celled organisms evolved into fish, which eventually left the water to venture onto land. Take a good look at a lungfish the next time you visit an aquarium or natural history museum. That will give you a fairly good idea of what the first species to leave the water looked like.

From fish came reptiles such as dinosaurs. The largest dinosaurs weighed in at around 100 tons, about the weight of a

> *It has become evident that the primary lesson of the study of evolution is that all evolution is coevolution: every organism is evolving in tandem with the organisms around it.*
> Kevin Kelly

> *It is not a monkey, not an ape and not a human, but it's a common ancestor of them all.*
> Russell L. Ciochon

modern streetcar. Some of these reptiles evolved into birds and mammals that eventually replaced dinosaurs as the prevailing forms of life on Earth about 65 million (65,000,000) years ago. Some of these mammals evolved into primates, from whence came the human species. The family tree shared by humans and apes begins approximately 4.4 million (4,400,000) years ago with the evolution of a common ancestor species that we commonly refer to today as the "missing link." At least that's the current thinking. An article written by Nick Wadhams for *National Geographic News* in August of 2007 reports that fossils from a newly discovered species dubbed *chororapithecus abyssinicus* may push the split between humans and apes back to 10 million years ago or even earlier, though this is uncertain just yet. I'm going to retain the presently accepted dates in this book with the full knowledge that I may need to revisit them later.

Humans share about 98.5% of our DNA with both chimpanzees and *bonobos* (pygmy chimpanzees), who are our closest living evolutionary relatives. The similarities of these three species are so great that an extraterrestrial zoologist would be forgiven for concluding that humans are a third species of chimpanzee. This zoologist may even be tempted to classify humans as *pan sapiens*, completely bypassing our self-imposed, self-indulgent, and separatist classification of *homo*.

Labels aside, humans and monkeys are too physically similar for any discerning observer to ignore. The few differences that do exist may owe much of their existence to a genetic mutation that occurred about 2.4 million (2,400,0000) years ago that caused human jaws to weaken. Weak jaws place less stress on the skull, allowing it—and the brain inside it—to expand in both size and capability. If this is true, then humans are not only monkeys, but diseased monkeys to boot!

> *Society does not consist of individuals but expresses the sum of interrelations, the relations within which these individuals stand.*
> Karl Marx

Losing one of our chief defenses against predators may have forced us to compensate by expanding our mental powers and replying on increasingly complex societal bonds. The story of human evolution is one of trading raw physical strength for unmatched mental ability. At least that's the short version.

The long version is that whatever emotional baggage you think your current or former significant other may have is only the tip of a huge evolutionary iceberg. Humans have what is

6 | The Natural Savage
Discovering the Human Animal

called a *triune brain* that consists of three layers: reptilian, mammalian (or *limbic*), and neomammalian (or *neocortex*). Each of us is literally carrying around millions upon millions of years of instincts, drives, and urges that live and interact inside our heads like a messy layer cake.

Have you ever gotten angry when a fellow driver cuts you off on the freeway? Ever pricked up your ears at hearing your flight called at a busy noisy airport despite having your nose buried in a book? If so, then you can thank your reptilian brain. Have you ever had "impure" thoughts about a member of the opposite (or same) sex? How about protective feelings toward a child or loved one? These are examples of your limbic brain at work. Finally, your ability to comprehend what you are reading at this moment comes courtesy of your neocortex.

I hold that it is true that dreams are faithful interpreters of our drives; but there is an art to sorting and understanding them.
Michel de Montaigne

The human brain and its built-in primal instincts and drives are roughly analogous to computer hardware. They form the nature side of the nature-versus-nurture equation. What we learn throughout our lives, especially during our formative years, is the software that we acquire through nurture. Our biology predisposes us to learn in certain ways and make certain interpretations, while experience teaches us when and how to invoke our primal drives. It's a circle of sorts.

Humans evolved from—and maintain both intimate connections to and relationships with—primates, monkeys, mammals, and all life on this planet. We cannot ever hope to truly understand human nature and our place in the grand scheme of things without looking at ourselves, not as separate forms of life, but as one of many animal species that inhabit the Earth.

The idea that humans evolved from apes therefore forms the second core assumption in this book. Some people find the idea of humans (*homo sapiens*) being a primate species alongside chimpanzees (*pan troglodyte*) and bonobos (*pan paniscus*) abhorrent. To them I can only say that the similarities between human and ape/mammal/animal behavior are too many to ignore, and that we can gain a greater understanding of humanity—on both an individual and collective basis—by observing our animal cousins.

Be a good animal, true to your animal instincts.
D.H. Lawrence

To Serve Man

Humans are such easy prey.
Stuart Gordon

This is the title of a fictional cookbook from a short story of the same name by Damon Wright that was later made into a *Twilight Zone* episode. As the episode ends, character Michael Chambers (played by actor Lloyd Bochner) laments that we will all be on the menu. What he doesn't tell us is that we're already on Nature's menu, and have been for millions of years.

It is both comforting and misleading to think of ourselves as sitting firmly atop the food chain. The reality is that humans are somewhere in the middle at best. Leopards and snakes routinely hunt both chimpanzees and bonobos, and are perfectly capable of dispatching an unarmed human. Conversely, chimpanzees, bonobos, and humans are all known to hunt, kill, and eat other animals.

Few modern humans need worry about being killed and eaten, but wild animals still do kill and eat people in remote corners of the world. Most monkey species, chimps and bonobos included, live with the threat of predation every day.

Birds also fit into this discussion. The harpy eagle feeds chiefly on monkeys and is compact enough to fly through the forest (as opposed to over it) in search of prey. Research performed by Lee Berger, a paleo-anthropologist at Johannesburg's University of Witwatersrand, indicates that our evolutionary ancestors were hunted by birds. Wounds inflicted in the skull of a two-million-year old *australopethicus africanus* (a human ancestor species) skull known as the Taung child skull are consistent with being killed by a predatory bird such as a large eagle. Studies of thousands of monkey skulls reveal similar injury patterns, such as holes and jagged cuts behind the eye sockets caused by *raptor* (bird of prey) attacks.

The lion cares less about being king of the beasts than about finding his dinner.
Mason Cooley

Predators have hunted us for millions of years, while the modern civilization that protects most of us from hungry animals has only been around for a few thousand years. Let's put this into some context: If the entirety of human evolution to date spans one hour, then modern humans have existed for less than 2 minutes, and civilization for less than 10 seconds. If we look at modern *homo sapiens* as having existed for one hour, then civilization has been around for less than 5 minutes. Why is this important? Because our brains have not had enough time to rewire themselves to reflect contemporary reality. Our prey instincts are alive and well, and are in large part responsible for our behavior.

Humans evolved in an environment filled with big cats on the ground, snakes in the trees, and eagles overhead. Until very recently, Mother Nature gave us no place to run to and no place to hide. Ours is a history bereft of refuges. The legacy of that reality haunts the core of our beings and lies at least partially behind everything we do.

What mythical creature do you get when you add a cat's legs and ears to an eagle's wings and a snake's body? Here's another hint: This creature breathes fire, the harnessing of which about 400,000 years ago is directly or indirectly responsible for virtually all aspects of modern life. It appears in just about every ancient, medieval, and contemporary culture on Earth. That's right: a dragon.

I believe that the mythical dragon is an embodiment of the predators that have feasted on our flesh since the dawn of our bloodline. How else can dragons, which combine our three primary predators with our single greatest technological advance, exist across civilizations that until far too recently had no way to communicate with each other?

An individual chimp, bonobo, or human doesn't stand much of a chance against a predator. That individual's odds get much better when it is part of a group. Monkeys and humans are social animals. Coincidence? I think not. A lone prey animal has a 100% chance of being the one selected for attack by a predator. A group of 25 prey animals reduces each individual's chances of being attacked to 4%. Even better, the individual unlucky enough to be attacked may benefit from the aid of its friends, thus increasing its odds even further. This can only occur when the group works as a cohesive unit, which can only result from an established leadership hierarchy. As any motivational speaker will tell you, there is no "I" in "team." Is it any wonder that humans develop serious physical and psychological illnesses if left alone too long? Is it any wonder that being ostracized from a group is one of the most traumatic things that can befall any person?

Prey animal survival depends on learning its predators' habits and developing behaviors designed to mitigate the threat of

Time goes, you say? Ah, no!
Alas, Time stays,
we go.
Henry Austin Dobson

For my own part, I would as soon be descended from that heroic little monkey or that old baboon as from a savage who delights to torture his enemies, offers up bloody sacrifices, practices infanticide without remorse, treats his wives like slaves, knows no decency, and is haunted by the grossest superstitions.
Charles Darwin

being killed and eaten. To take this a step further, a prey animal's entire life revolves around avoiding being killed and eaten. As for the predators? Any nature program will show you lions and eagles sunning themselves in plain view of everyone, and why not? With rare exception, nothing is going to kill and eat them!

To take this concept one step even further, I believe that the need to band together to avoid predators gave rise to tribes and nations. Yes, there is also the need to secure and defend territory for food and shelter, but I believe this to be secondary in importance. The way I see it, it's hard to be hungry or to exercise any of our other instincts and drives when you're dead.

The idea that humans evolved from—and remain—prey animals is the third core assumption in this book. This may be a tough concept to swallow, because it's more than a little humbling to think of ourselves as anything less than the paragon on animals, the absolute pinnacle of evolution. But you know, a little humility may be a good thing, especially if it helps us gain a deeper understanding of who we are and how we can use that knowledge to benefit all of humankind.

Animals, in their generation, are wiser than the sons of men; but their wisdom is confined to a few particulars, and lies in a very narrow compass.
Joseph Addison

Creation

Did God (or a god) create the Universe or is it the by-product of random quantum phenomena? If the Universe is the result of an intelligent design, is God directly involved in all of its many goings on, or merely the force that started the cosmic dominoes falling? Is there life after death, or is our brief life span all we get? If there is life after death, then does life on Earth represent our first phase of existence, or have we had past lives? Should I affix a little metal fish to the back of my car and, if so, should it say "Jesus," "Darwin," or "Gefilte"? What if 42 really is the meaning of life, and what if the hokey pokey really is what it's all about?

All levity aside, the question of evolution versus creation really is deadly serious. If this chapter hasn't established beyond doubt that I believe in evolution, then nothing will. I happen to believe that God (in a generic sense as opposed to the god

God gave us intelligence to uncover the wonders of nature. Without the gift, nothing is possible.
James Clavell

of any religion) created evolution, a concept I explore in depth in *The Divine Savage*.

The good news is that you need not believe me in order to read this book and gain some useful insight from it. If humans are nothing but biological creatures whose brains are the source of our consciousness (which I strongly doubt), then we are constrained by our brain's abilities and limitations. If humans possess a soul or spirit with our brains being mere conduits for consciousness (as I strongly suspect), then we remain constrained by the same limits.

One cannot obtain a high-definition TV picture from a non-HDTV set, no matter how good or strong the signal is. The same logic applies to the topics covered in this book, which discusses *what* humans do and the biological reasons *why* we do it, without needing to explore *where* all of this ultimately came from. This is my fourth and final core assumption.

I'm astounded by people who want to 'know' the universe when it's hard enough to find your way around Chinatown.
Woody Allen

In the beginning the Universe was created. This has made a lot of people very angry and has been widely regarded as a bad move.
Douglas Adams

Chapter 2

Reducto ad Absurdum

The fixity of a habit is generally in direct proportion to its absurdity.
Marcel Proust

What an amazingly complex society we live in! Billions of people in countless cities and towns do millions of things from the mundane to the marvelous, and busy themselves in thousands of different industries every day. Modern humans are a far cry from the small loose-knit tribes of hunter-gatherers we evolved from, whose last vestiges still inhabit the remotest regions of the Earth, such as the rainforests of Brazil and lush valleys in Papua New Guinea. Or are we?

Life at a Glance

Boiling away the seeming complexity of any present or past civilization reveals that all of human existence revolves around six absurdly simple core functions:

- Predator avoidance
- Group status
- Sustenance
- Shelter
- Reproduction
- Death

By learning to discover and value our ordinariness, we nurture a friendliness toward ourselves and the world that is the essence of a healthy soul.
Thomas Moore

For all of its seeming intricacy—and despite the many devices humans use to distance themselves from the truth—modern life remains just as absurdly simple in essence as it has been since the dawn of time.

Every animal species, human and otherwise, performs the same six core functions. This alone blurs the distinction so many of us want to make between humans and animals. To quote Charles Darwin, the difference in mind between man and the higher animals, great as it is, certainly is one of degree and not of kind. Indeed, a careful examination of the many similarities between humans and animals forces one to conclude that the degree of separation is very small indeed.

> *Life is what happens to you while you're busy making other plans.*
> John Lennon

The Six Functions of Life

Let's follow a typical human life from birth to death to illustrate how the six functions of predator avoidance, group status, sustenance, shelter, reproduction, and death come into play throughout our lifetimes. Later chapters will delve into each of the six basic functions in much greater detail.

Predator Avoidance

From an evolutionary standpoint, the sole purpose of life is living long enough to pass on one's genetic material. The vast majority of animals on this planet occupy the low to middle tiers of the food chain, meaning that most animals have to worry about being killed and eaten. It is the rare animal that has nothing to fear from predation. Since predation renders all else moot, it stands to reason that a prey animal's survival depends on learning how its predators operate, and then avoiding them at all cost. For example, few gophers will venture from their burrows at high noon when there are hungry hawks overhead.

Some behaviors can be classified as innate, or instinctual. For example, many animals (humans included) have a fear of falling, and will avoid cliffs or anything that looks like a cliff. Numerous "visual cliff" experiments with a transparent surface placed above an obvious drop-off conclude that many creature simply will not cross the gap, their assured safety notwithstanding. Other behavioral traits such as whether the ani-

> *You think Nature is some Disney movie? Nature is a killer. Nature is a bitch. It's feeding time out there 24 hours a day, every step that you take is a gamble with death. If it isn't getting hit with lightning today, it's an earthquake tomorrow or some deer tick carrying Lyme disease. Either way, you're ending up on the wrong end of the food chain.*
> Jeff Melvoin

Chapter 2
Reducto ad Absurdum

mal is diurnal or nocturnal, whether or not it hibernates, etc. are also innate.

That still leaves plenty of room for learning. All parents of any species who raise their young do so to impart survival skills that begin with predator avoidance and continue on to appropriate group behavior, finding food, and all of the many things the child must know in order to live and have a decent chance at reproduction.

Humans evolved from prey animals and, but for technology, would be in the middle of the food chain at best. It is true that few of us alive today face predation at any point in our lives, but animals still do kill and eat people every year. Even more importantly, our brains remain those of prey animals because modern civilization has only existed for a few thousand years, while predators (chiefly big cats, snakes, and birds of prey) have hunted us for millions of years. Our brains just haven't had time to rewire themselves to reflect our modern reality.

From the start, human infants interpret everything they experience as survival instructions. This concept is extremely important, as I previously described in *The Enlightened Savage*, and as I will discuss in greater detail in Chapters 5 through 9.

Group Status

Many animals come together in groups to protect against predators, and both apes and humans most certainly fall into this category. Remember that ape and human predators are big cats, snakes, and birds of prey such as Panama's harpy eagle. A lone ape or human doesn't stand much of a chance against a predator, and is therefore at a fatal disadvantage. But if these apes and humans can band together...

One prey animal has a 100% chance of being attacked by a passing hungry predator. A group of 25 such animals drops the odds of any one individual being attacked only 4% assuming that all else is equal. Even better, a group of animals may be able to fight off the predator and emerge unscathed.

On the other side of the prey/predator equation, hunting in packs allows the hunters to take on much larger prey, such as a pack of hyenas attacking a wildebeest. Groups of human hunters have successfully bagged mammoths, elephants, buffalo,

---

A good way to avoid predators is to taste terrible.
Unknown

When spider webs unite they can tie up a lion.
African Proverb

and plenty of other large animal species whose individuals could easily outrun or outfight a lone hunter.

Groups make lots of evolutionary sense for animals that can't outrun or outfight their predators, while also allowing animals to take on prey that is beyond any single hunter's abilities; however, the only way groups can work is if they have solid leadership. Otherwise, the group is nothing but a bunch of individuals doing their own thing with no cohesion or common purpose whatsoever. Hierarchies are therefore common features of social animals, humans included.

Group leaders (alphas) typically enjoy privileged access to resources up to—and sometimes including—reproductive rights. In theory, those individuals in leadership positions have triumphed over weaker opponents in addition to successfully facing life's many challenges, and are therefore the ideal candidates for passing along the almighty genes.

Our evolutionary cousins, the chimpanzees and bonobos, live in hierarchical groups complete with alliances and political maneuvering for position. Sound familiar? Every single group you belong to—from your knitting circle to your nation—has leaders, followers, explicit and/or implicit rules, some means of enforcing those rules, and some process for selecting leaders. If you're thinking that your group is totally informal with no real leader, think again.

Nonconformists travel as a rule in bunches. You rarely find a nonconformist who goes it alone. And woe to him inside a nonconformist clique who does not conform with nonconformity.
Eric Hoffer

Learning how the group operates, and how to fit in and gain status, is a critical subject for many animals. Human children and adolescents invest seemingly inordinate amounts of energy trying to fit into and be part of the "in" group as they seek and jockey for position, popularity, and rank. The pariahs and outcasts at the bottom of the heap tend to be depressed, which may be an evolved response to losing rank or status. Depression can suppress the urge to keep fighting a losing battle, thus enhancing group harmony and reducing injury to the loser.

If you've been unlucky enough to be fired or expelled from a group, then you probably know what I mean when I assert that expulsion from a group may well be the single most traumatic event that can befall a person. Being the one doing the expelling can also be very traumatic. Why? Because humans evolved from—and remain—prey animals, whose survival depends on belonging to a group. At a subconscious level, your brain doesn't know that you hated your job and are better off being let go. All it knows is the pain of rejection that stems from the primal fear of being outcast and alone, and thus extremely exposed and vulnerable.

A large percentage of a child's time is spent learning how to interact with other members of its society, both to gain protection from predators, and to enhance its chances of successful reproduction. Chapters 10 through 14 deal with status in more detail.

Sustenance

> *Food is an important part of a balanced diet.*
> Fran Lebowitz

Whatever did we do in the days before supermarkets brought our food to us in convenient heat-and-serve portions? The same thing most animals occupy most of their time with day in and day out: Looking for food.

If you're a koala whose diet consists of eucalyptus leaves, then your decision about what to eat becomes simplicity itself: If it looks, smells, and tastes like a eucalyptus leaf, then chances are excellent that you've just found lunch. But what if you're an omnivore capable of eating much of the mind-boggling array of fish, fruits, vegetables, aquatic plants, mammals, eggs, tubers, roots, and fungi that occupy virtually all life-bearing areas on Earth? Even better, what if certain parts of certain plants are only palatable when cooked, or if certain parts of certain fish contain deadly toxins? What happens when some of the animals being hunted are perfectly capable of maiming or killing their hunters, should they turn and fight instead of dying quietly? Some vegetables, such as the cassava root (do you like tapioca pudding?), contain poisons that are neutralized by cooking, and some shellfish are toxic at certain times of the year. These are just a few examples of the many dangers lurking in our potential food supply. Welcome to what Michael Pollan calls the *omnivore's dilemma*.

> *A day's work is a day's work, neither more nor less, and the man who does it needs a day's sustenance, a night's repose and due leisure, whether he be painter or ploughman.*
> Ralph Waldo Emerson

In his book by the same name, Mr. Pollan argues that an omnivorous diet places huge demands on our brains. After all, how else can we possibly remember which foods are edible under which circumstances? Common wisdom tells one to take a small bite of any unfamiliar food, wait several hours, and then try progressively larger doses at equally long intervals. Experiments with rats show that rats remember what makes them sick and will not ingest more of something with ill effects. If you've ever suffered food poisoning, then it's a safe bet that you thought long and hard before again partaking of the same thing again.

Parents pass along knowledge about what's good to eat and how to find and prepare food to their young, and members of a group or tribe also share knowledge among themselves. This requires communication. Language may well have evolved at least in part to help us store and pass along knowledge about what's good to eat. If you hold down a job or own a business, then much of your daily efforts are going into finding food, or at least the medium of exchange (money) used to obtain food. Chapters 15 through 17 talk about food in more detail.

Reminds me of my safari in Africa. Somebody forgot the corkscrew and for several days we had to live on nothing but food and water.
W.C. Fields

Shelter

From caves to luxury condominiums, thatched huts to cardboard boxes and ski chalets, shelter is a fundamental need for humans and countless other species that invest significant energy finding, constructing, and maintaining various forms of shelter. Seeking or maintaining shelter occupies a significant chunk of our time, whether we're at work earning money to make the rent or mortgage payment, or weeding the yard. Indeed, many of us spend 40 hours or more per week toiling away at seemingly unrelated assignments in order to afford food and shelter. See Chapter 18 for a discussion about shelter.

The poor on the borderline of starvation live purposeful lives. To be engaged in a desperate struggle for food and shelter is to be wholly free from a sense of futility.
Eric Hoffer

Reproduction

Everything from avoiding prey to fitting into the social hierarchy and securing food and shelter revolves around—and supports—the one life mission evolution has dictated for us: reproduction. The human race is the only species currently known to understand the connection between copulation and subsequent birth, and has constructed elaborate legal and moral codes that attempt to regulate sexual behavior, from mate selection to codifying what is and isn't appropriate sexual conduct. Did you know that sex in any position other than the missionary position is illegal in Washington, D.C?

Life is a sexually transmitted disease.
R. D. Laing

Here again, the similarities between humans and animals are striking. Animals engage in recreational sex, orgies, homosexuality, incest, adultery, polygamy, and just about every other sexual behavior imaginable. Don't believe me? Chapters 19 through 23 discuss reproduction in more detail.

Death

When we put off our clothes at night, we have a fit occasion to consider, that we must strip nearer one of these days, and put off, not our clothes only, but the body that wears them too.
John Flavel

The sad fact is that evolution does not care about any one individual; survival of the species is what counts, and this is why death is the inevitable end of life as we understand it for all living things.

Plants rely on the decomposed remains of other plants, supplemented by nutrients from animal waste (itself a by-product of dead plants and/or animals), or from the decaying corpses of the animals themselves for nutrition. Herbivorous animals kill and eat the plants, and are themselves killed by carnivorous animals. Animal excrement and remains nourish the plants, and the great circle of life is complete. Think about every single thing you eat, and you'll soon realize that your very life depends on killing other living things every single day.

Beyond nourishment for others, death makes room for new life to flourish. I once read that suspending death for just a few days would bury the planet under several feet of insects. Approximately 112 humans die per minute. How many seconds did it take you to read this paragraph? Multiply that number by two, and that's roughly how many people have died in the same amount of time.

One death is a tragedy. A million deaths is a statistic.
Josef Stalin

Finally—and most importantly—death allows life to evolve. Sexual reproduction mixes genes, and creates new mutations that allows the species to evolve. Individuals that die out without reproducing curtail genes that aren't suited for the species's current environment, freeing up more room and resources for those whose genes are more favorable. Those who do survive long enough to reproduce pass on their genes to the next generation. Over time, unsuitable genes are weeded out through death. You are alive to read this book thanks to an unbroken chain of survival that extends back through billions of years of evolution to the first single-celled organism on Earth. Changing environments and evolution of other species alongside our own evolutionary ancestors created pressures that favored the genetic lineage that eventually gave rise to *homo sapiens* and from there to Anthony Hernandez and yourself.

Left to its own devices without medicine, the natural life span lasts just as long as is needed to maximize the odds that some or all of an individual's genes will be passed on, no more, no less. A number of factors come into play when determining the ideal lifesaving for any species, including:

- Wear and tear. All things wear with age, and bodies are no exception. Just like an aging car, repairing a living thing becomes more and more costly and laborious until reaching the point where it makes more sense to simply start over. Odds of injury and disease and the cost to repair versus replace factor into this equation. For example, it makes little sense not to repair a simple cut or bone fracture in monkeys and humans. The cost of regrowing a severed limb, however, is just not worth it because of the greatly increased odds of further injury or predation caused by loss of mobility.

- Average reproductive rate. Insects that release thousands of eggs that mature in a few days don't need long life spans. By contrast, apes and humans whose offspring require months of gestation followed by years of parenting require much longer life spans in which to be able to accomplish these tasks.

- Predation risk. How many offspring can expect to reach adulthood? The many thousands of insect eggs and larvae produced by some species are mostly eaten long before becoming fertile themselves, but enough survive to release thousands more young. Animals born immature or otherwise helpless need parental protection to survive, and this in turn requires a longer life span.

- Rearing kin. There are many cases in monkey and ape groups where siblings, relatives, and other adults help rear the young. This does not directly pass on a caretaker's genes, but does help ensure that the caretaker's bloodline will persevere since all relatives carry some portion of her or his genes with their own.

Add all of these factors up, and you'll see that the natural life span of any species is just long enough to provide the opti-

---

Nothing in the entire universe ever perishes, believe me, but things vary, and adopt a new form. The phrase "being born" is used for beginning to be something different from what one was before, while "dying" means ceasing to be the same. Though this thing may pass into that, and that into this, yet the sums of things remains unchanged.
Ovid

---

mum cost/benefit ratio for that species. For humans and some domestic animals, modern medicine has trumped nature and given us unnaturally long lifetimes. Thankfully, contemporary life provides many ways of carrying out life's six primary functions, thereby extending most people's "useful" life span. Still, it should come as no surprise that suicide rates are higher among the elderly, who presumably don't want to impose a burden on their families or on society. Life is just as long as it needs to be. Any longer would be a waste, and nature wastes nothing.

Death is unavoidable. This begs the question of whether life is purely biological or if there is another plane of existence beyond this one. In other words, is death the end, or merely a new beginning?

On the plus side, death is one of the few things that can be done just as easily lying down.
Woody Allen

20 | *The Natural Savage*
Discovering the Human Animal

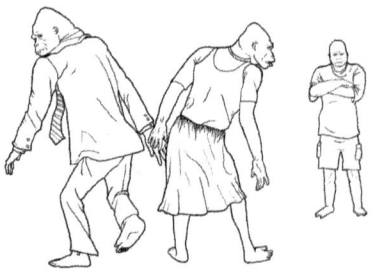

Chapter 3

The Other Chimpanzees

> *A loving dog-owner can tell you more about animal awareness than some laboratory behaviorists.*
> George Shaller

In *The Third Chimpanzee*, Jared Diamond speculates that an alien zoologist landing on Earth would see three species of chimpanzees inhabiting our planet: *pan troglodytes* (chimpanzees), *pan paniscus* (pygmy chimpanzees, also called bonobos), and *homo sapiens* (humans). This zoologist might even go so far as to classify humans as *pan sapiens* or to classify the three chimpanzee species as *homo*, resulting in *homo troglodytes*, *homo paniscus*, and *homo sapiens*, respectively. In fact, as we saw in Chapter 1, an article written by John Pickrell for *National Geographic News* in May of 2003 makes a strong case for the latter. Let's start looking at why this might not be a bad idea.

Tearing Down the Wall

> *Life is just a mirror, and what you see out there, you must first see inside of you.*
> Wally 'Famous' Amos

Nobody knows when humans began to think of ourselves as magically above or beyond animals, but that self-imposed distinction has manifested itself in laws, moral codes, traditional psychology, and religious canons that both contravene human nature and maintain the philosophical wall separating us from the rest of the animal kingdom. One need look no farther than Genesis 1:26 and the Koran Sura 11:6 (which references the Genesis story) to see this wall in the form of God giving humans dominion over the Earth and all its creatures. And for what? The world of today is rife with suffering, poverty, violence, and pollution that may spell the end of all life on the one life-bearing planet known to exist in the entire vastness of the Universe.

Proponents of the idea that humans are different than animals argue that to remove the wall separating us from nature would devalue humanity and its achievements, rendering us mere dumb soulless brutes, and stripping us of all that makes us human. I believe the exact opposite. I believe that acknowledging ourselves as animals would reconnect us to the grand continuum of life that extends from the simplest bacteria to the most complex creatures. Rebuilding this connection would remind us of the obligations wrought by our developed intellect and technology to be stewards of the wondrous planet we live on. It would help us reconstruct society and its laws and customs to fit our evolved nature, thus reducing much of our stress and discontent. Free to be the animals we are, how much more rewarding and fulfilling life might be!

There are two kinds of people in the world, those who believe there are two kinds of people in the world and those who don't.
Robert Benchley

What if, as John Lennon dared us to imagine, there is no Heaven above us, no God? What if all we are is the product of random change over billions of years, mere bags of meat whose perception of existence begins at birth and ends at death? If so, then every passing second is gone forever, and we can't afford not to wring every last ounce of fulfillment out of every one of our precious seconds. And if there is a God? I can think of no better way to exalt the Creator than to understand and live in accordance with what It has wrought. Either way, we owe it to ourselves to study evolution and the portion of human nature that consists of evolved instincts and drives.

Common Heritage

People in restaurants use menus to peruse all available options. They then select those choices that best suit their current tastes. Requests for changes or substitutions tend to be rare, and are also constrained by what the restaurant has on hand. I've never heard of a diner bringing food to a restaurant for preparation. Quite the contrary: the diners select which menu items best suit their current circumstances, eschewing other possible choices. Evolution works the same way: Those individuals and species best able to thrive in the current environment (circumstances) have the best odds of reproducing.

I am convinced that the world is not a mere bog in which men and women trample themselves in the mire and die. Something magnificent is taking place here amid the cruelties and tragedies, and the supreme challenge to intelligence is that of making the noblest and best in our curious heritage prevail.
Charles Austin Beard

> *The theory of evolution by cumulative natural selection is the only theory we know of that is in principle capable of explaining the existence of organized complexity.*
> Richard Dawkins

It is important to note that individuals in a given species are more or less adapted to the current environment. Every individual's genes specify greater or lesser capacities for certain traits that are either favorable for conditions, or not. Genes provide the capacity for behavior, but they neither explain nor justify every action. A sudden change, such as an ice age or meteor strike, will wipe out all life that is incapable of surviving the new conditions, leaving behind those with the good fortune to possess traits suitable for the new conditions. The same thing happens with modern antibacterial products that claim to kill up to 99.9% of germs.

What the advertisements for these products don't tell you is that surviving 0.1% possess whatever resistance is needed to survive the chemical onslaught. Repeated applications of the same chemical will breed resistant germs until the product ceases to be effective, and your home becomes full of super bugs that may also have the capacity to do you more harm than the germs you were so afraid of in the first place. Overeager applications of medical antibiotics and patients who stop taking their medicine when they feel better are rapidly breeding drug-resistant diseases that are wreaking havoc on those who contract them. This is a prime contemporary example of natural selection favoring certain traits above others from among the available options.

The natural selection process has shaped countless millions of species over billions of years. It is therefore ludicrous to think that humans are the one exception to this history. The proof of this assertion lies in the fact that humans share 98.5% of our DNA with both chimpanzees and bonobos, who in turn share 99.3% of their DNA with each other. By comparison, humans share 97.7% of our DNA with gorillas.

> *The real voyage of discovery consists not in seeking new landscapes but in having new eyes.*
> Marcel Proust

The Great Divide

What caused the evolutionary split between apes and humans? The answer seems to lie in climate change. The savanna theory holds that the vast tropical African forests dwindled, leaving large expanses of grassland. The woodland mosaic theory speculates that our *australopithecus* ancestors may have survived in patchy forests interspersed with grass. Finally, the variability theory says that repeated climate changes may have forced our ancestors to survive under varying conditions.

Leaving aside the question of how it happened, human ancestors adapted to life on the ground, which included walking upright (among many other adaptations). Still, our behavior remained remarkably ape-like for millions of years.

From Ape to Man

Climate change may have been a mechanism that finally separated ape and humans from a behavioral standpoint, as well as facilitating the physical changes that distinguish us from apes. Faster sexual maturity brought on by the need to reproduce under less-than-favorable conditions could result in *juvenilization*. Juveniles tend to have larger heads relative to their adult counterparts. The problem with this is that juveniles tend to be smaller, which could limit their ability to run and hunt, so other genes favoring size would come in handy. At some point, however, larger heads become problematic when they can no longer fit through the birth canal. Slower development and immature babies solve this problem. Combine that with the mutation that may have caused jaws to shrink and thus allowed brains to expand, and the stage seems set for humans.

I know this—a man got to do what he got to do.
John Steinbeck

Anatomy

This section touches on some of the key differences between humans and chimpanzees, and offers some intriguing ideas why some of our shared traits may have evolved as they did.

Anatomy is destiny.
Sigmund Freud

Limbs

Chimpanzee arms are much longer than their human counterparts, perfect for arboreal life, because it's much easier to hang underneath small branches than to try to balance atop them. Large arms accommodate large muscles, which are also useful for arboreal life. By contrast, their legs are shorter than ours. Chimpanzee fingers and toes are much longer than ours, which makes them more suitable for grasping branches. Chimpanzee feet are *prehensile*, meaning that they too can wrap around branches and hold on tight.

Chimpanzee pelvises are aligned with the spine, unlike human pelvises that are rotated to adapt for bipedal walking, and that also provide an excellent support basket for our internal

He that climbs the tall tree has won right to the fruit.
Sir Walter Scott

organs to keep them from sagging downward. Human legs are designed for bipedal motion, from the angle at which our femurs connect to our pelvises, to the shock-absorbing arches in our feet. Our largest muscles are in our legs, another adaptation for walking upright.

Bonobos are both smaller and more *gracile* (slimmer and less robust) than chimpanzees in general, and they are better adapted for bipedal motion than chimpanzees, thanks to their smaller pelvises and leg bones that are better adapted to upright walking along the same lines as humans.

Skulls

Cabbage: A familiar kitchen-garden vegetable about as large and wise as a man's head.
Ambrose Bierce

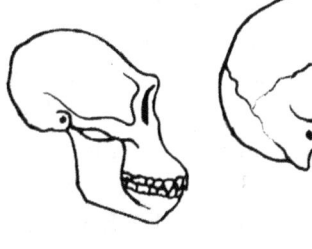

Chimpanzee skulls feature heavy jaws and teeth suitable for chewing tough fibrous foods, such as bark and nuts. The brain case is a small (relative to humans) cavity that sits well behind the teeth. Prominent brow ridges are evident. It's as if the entire skull has been reinforced to withstand heavy stress. The much weaker human jaw places much less stress on the skull, which arguably permitted the brain to grow upward and (most importantly) forward. We'll discuss the importance of frontal brain lobes in Chapter 4.

Vision

I started concentrating so hard on my vision that I lost sight.
Robin Green

Humans and chimpanzees have *trichromatic* vision, meaning that we see red, green, and blue (from which all secondary and other colors are derived). Apes depend on fruit and vegetables for much of their diets. Keen color vision is a great advantage, because it makes finding fruits and vegetables much easier. It is also very useful for distinguishing branches.

I have already asserted that humans and our evolutionary cousins are much more prey animals than predators. Some have challenged this assertion because human and ape eyes are forward-facing like most predator animals, which gives us the 3D vision needed to accurately track and catch prey. Many prey animals' eyes are located on the sides of their heads, the better to see danger sneaking up behind them.

This argument makes sense until one remembers that apes live in trees. Having done my fair share of mountaineering and rock-climbing in my younger days, I can say with absolute certainty that 3D vision is a huge help. How else could an ape know exactly how far to reach to grasp a branch or accurately judge its motion through the trees? For that matter, how could an ape judge whether a branch is large enough to support its weight? One mistake could prove fatal. 3D vision is essential to arboreal life. It also doesn't hurt when hunting.

Consciousness

Chimpanzees and bonobos are some of the few animals known to possess *self-consciousness*, defined as a sense of self as an entity separate from one's environment. Human children require about a year to develop this sense. I learned this firsthand by observing my son Logan. He could respond to questions about the location of people and objects in the room, but gave me a blank look whenever I asked "Where's Logan?" One day, he pointed at himself. That's when I knew that he had developed his own sense of self.

Life is consciousness.
Emmet Fox

Researchers proved that chimpanzees possess self-awareness by painting a colored dot on their foreheads while they slept, and then showing them a mirror upon waking. The apes responded by rubbing their foreheads. Animals that fail this test apparently do not have a sense of self (although this is debatable).

Why should a sense of self be so useful for chimpanzees and humans when so many other species apparently get by just fine without it? Once again, our arboreal ancestry provides a clue: When swinging through the forest, having a sense of self distinct from one's surroundings is a valuable tool, because it allows one to better plan one's actions and foresee the consequences of any mistakes.

Diet

Chimpanzee and bonobo diets are very alike. As you'll soon see, human eating habits diverged when we began eating more meat.

Water is the most neglected nutrient in your diet but one of the most vital.
Kelly Barton

Chapter 3
The Other Chimpanzees

Chimpanzee

Chimpanzees primarily eat ripe fruit and also enjoy raw leaves, all eaten right where they found it. Seeds, bark, pith, and flowers are important supplements during the dry season. They also eat termites, peeling the bark off long narrow twigs that they then poke into termite mounds to "fish" for the insects. Females eat about twice as many termites as males. Chimpanzees also eat meat, primarily monkeys (including occasional cannibalism), pigs, and antelopes. They have even been observed to make and use spears that they thrust into burrows to hunt bush babies, with a success rate significantly higher than most other hunting animals. Still, meat accounts for only a small part of the chimpanzee diet.

> *Food is the most primitive form of comfort.*
> Sheila Graham

Bonobo

Bonobos have the same overall diet as chimpanzees, except that meat is a smaller part of their diet, because the pith they eat has abundant protein.

Human

Much of the human diet resembles that of chimpanzees and bonobos, in that we eat an astonishing variety of fruits, vegetables, and leaves. We also eat a wide variety of grains. As for fish and animals, I've watched enough exotic cooking shows (ever see the Japanese classic *Iron Chef*?) to suspect that few species are off the menu.

The expanded percentage of meat in the human diet made for an interesting change in our eating habits. Whereas herbivores spend much of their time foraging and snacking, carnivores tend to eat infrequent big meals—a natural consequence of the fact that fresh meat is nowhere near as ubiquitous as fruit and other plants. This probably goes a long way to explaining our normal routine of breakfast, lunch, and dinner. Interestingly enough, the idea of frequent small snacks is enjoying renewed interest in people as a means of combating the growing obesity epidemic.

> *Thought: Why does man kill? He kills for food. And not only food: frequently there must be a beverage*
> Woody Allen

Range

Common chimpanzees range across equatorial Africa, from Senegal in the west, through Zaire (Democratic Republic of Congo) north of the Congo River, to Tanzania in the east. They inhabit tropical forests and grasslands with abundant fruiting trees. Bonobos only live in a small area of Zaire south of the Congo River.

A man's homeland is wherever he prospers.
Aristophanes

Society

Chimpanzee and bonobo societies are strikingly similar in some ways, and just as strikingly different in other ways. As you read, look for the ways in which both societies compare with our own. Chapters 10 through 14 discuss status within society in more detail.

Society, my dear, is like salt water, good to swim in but hard to swallow.
Arthur Stringer

Chimpanzee

Chimpanzee males are very status-conscious and readily primed for a fight. Raised hair is a clear sign of tension. Aggressive chimp males can charge (with or without branch in hand), throw rocks, and attack anyone in their path. Submissive individuals will emit panting grunts or scream in fear. Friendly encounters may include hugging and kissing.

Males jockey for status within their groups, and can form alliances that both help individuals rise to power and secure standing and favors for those aiding him. Male chimpanzees wander from group to group, while females tend to remain in the same group, thus helping ensure genetic diversity by preventing inbreeding.

Grooming is extremely important. This social activity helps keeps chimps clean and free of parasites. It also relaxes the ape being groomed, and is thus an effective means of reducing tension—the chimpanzee equivalent of friendly small talk. Females mostly groom family members, while males mostly groom other males.

Females in heat attract groups of males, and may mate with them or may be taken by a high-ranking male. Rarely, female-female or male-male mountings occur as a means of reassurance.

Chimpanzees are *diurnal*, meaning they sleep at night. Every evening, they build fresh nests in trees using leaves and branches. They may occasionally reuse old nests, and occasionally wake to feed during the night on moonlit nights.

Bonobo

Bonobos live in large groups that split up into smaller groups to forage during the day. Groups keep their distance, but rarely conflict when they chance upon each other. Like chimpanzees, they sleep in arboreal nests. Unlike most primates, bonobos are known to reuse the same nests.

While chimpanzee societies are male-dominated, bonobos are decidedly female-oriented. Males remain in the same groups and form lasting bonds with their mothers, in part because of their extremely slow development. As part of this arrangement, mothers help their sons gain status, which affords them more mating opportunities, which boosts the mother's chances of passing on her genes and winning the great game of life.

Order is not pressure which is imposed on society from without, but an equilibrium which is set up from within.
Jose Ortega y Gasset

The needs of society determine its ethics.
Maya Angelou

Female bonobos travel from group to group to avoid inbreeding. Related females (siblings, mother/daughter, etc.) do not join the same group, another means of preserving genetic diversity. Females wield the real power within the group, in a role reversal from chimpanzees. They also form very strong bonds despite not being related. Still, loose male hierarchies do exist, and those of higher rank have greater mating opportunities.

Chimpanzees often resolve conflicts by fighting, which can progress to the point of warfare between groups. By contrast, bonobos rarely engage in violence (including hunting). They instead opt for reconciliation, which very often consists of sexual contact. Bonobos have been called the apes from Venus, and with good reason. Male-female, female-female, and male-male sex is common among bonobos. Female-female sex is particularly important because of the tight bonds between females in a group. Female-male sex can be used a tool to obtain food or other favors—and you thought humans were unique in this regard! The heavy prevalence of sex within bonobo groups means that bonobos don't know their fathers.

Human

Humans societies borrow from both chimpanzees and bonobos. Males are usually dominant, and both males and females travel from place to place. Individuals may resolve conflicts peacefully, but may also resort to verbal and physical violence. We also have lots of sex in every conceivable position and combination, much of it for purely recreational purposes. We'll examine this in more detail throughout this book.

Tools

Both chimpanzees and bonobos use tools. Chimpanzees use branches in threat displays. They also peel bark off twigs that they use to "fish" for termites. To crack nuts, they will place a seed on a stump or suitable rock and use a branch to strike the nut with astonishing precision. Branches are sometimes used as weapons during chimpanzee wars.

Bonobos also use tools, but to a much lesser extent that chimpanzees, because they don't hunt as a rule. They will mash the

If the only tool you have is a hammer, you tend to see every problem as a nail.
Abraham Maslow

end of a stick for use soaking up liquid to drink. The most amazing example of tool usage among primates is Kanzi, a bonobo who has learned to make and use stone flake tools to perform complex tasks, such as cutting ropes to access food inside a secured container.

Language

If the English language made any sense, a catastrophe would be an apostrophe with fur.
Doug Larson

Chimpanzees have a range of calls used for various purposes. One common call is a loud rising hoot that announces an individual or groups' presence. Excited "barks" announce a food source. Chimps being attacked often scream repeatedly. Other calls alert to danger or screams. The previously mentioned "pant grunt" indicates submission. Hand and foot gestures are also used. Bonobos use a similar system of calls and gestures. They will also solicit sex as a means of conflict resolution, obtaining food, cementing social bonds, and more.

Humans have trained both chimpanzees and bonobos to recognize spoken words, sign language, and languages composed of pictographic symbols. Does this mean that they possess language in the wild or an aptitude for learning true language? Beyond toolmaking, Kanzi has exhibited a breathtaking knack for communication. Neither chimpanzees nor bonobos have language in the human sense under natural conditions, but then again we may be defining language in *anthropomorphic* (human-centric) terms that exclude other animals. Seen from the animals' perspective, I find it hard to imagine that they don't possess language that is eminently suited to the needs of each species. I'll talk more about animal language in Chapter 5.

Reproduction

Nobody will ever win the Battle of the Sexes. There's just too much fraternizing with the enemy.
Henry Kissinger

Let's take a brief look at how our evolutionary cousins reproduce. As you read, take note of the many similarities between chimpanzee, bonobo, and human reproductive behaviors.

Chimpanzees

Both males and females mate with multiple partners; however, males sometimes limit access to females, either by force, or by taking the female on a *consortship* that can last several months.

A female chimpanzee's genitals swell and turn bright pink for about seven days out of her 36-day cycle. This provides a clear advertisement of fertility, and is the time when the female is most sexually receptive. Females in heat often receive preferential treatment that can include peaceful interactions with unknown males.

Infants and juveniles sometimes copulate with a female in the early stages of heat for practice. One or more males may copulate with her nearer the peak of her cycle. An ovulating female may be taken over by dominant males, who prevent others from mating, often by attacking her if she shows interest in other males. Attacking the female is less risky than attacking other high-ranking males, and often gets the point across effectively. This reduces the potential father pool. Given a choice, males tend to prefer females who are older, calm, and relatively unknown. Females may tend to prefer males who lavish them with more friendly attention. Both genders often exhibit an incest taboo.

Females give birth to a 4.5-pound infant (occasionally twins) every three to six years, following a gestation of about seven and a half months. Infants cling to their mothers' chests for the first three to six months, after which they usually ride on her back until weaning at roughly four years of age. At this point, young chimpanzees can survive on their own, but typically remain with their mothers until they mature at about age 10. Puberty occurs at about age seven. In Tanzania, females typically birth their first young when they are about 15 years old. Females do most of the parenting work, and children bond with their mothers for life. Sibling bonds are also strong, with the older children helping care for the young. Males play with young chimps and offer protection against strangers.

Bonobos

This section is shorter than the one on chimpanzees because we know much less about bonobos than the former. What we do know is that bonobos are the second most promiscuous primates on Earth, next to humans. Much of their sex occurs for recreational or social bonding purposes, with reproduction seeming to be almost a by-product of these other activities. The key difference between bonobos and humans in this

In America sex is an obsession, in other parts of the world it is a fact.
Marlene Dietrich

Well, you're either lovers or you're wanting to be lovers or you're trying not to be lovers so you can be friends, but any way you look at it, sex is always looming in the picture like a shadow, like an undertow.
Diane Frolov and Andrew Schneider

Chapter 3
The Other Chimpanzees

regard is that bonobos haven't figured out that children are the natural consequences of their activities.

Female estrus lasts 10-20 days, much longer than in chimpanzees, and most mating occurs at the peak of the cycle. Estrus seemingly resumes within one year of giving birth, though the female probably remains infertile for a longer period. A mother will give birth roughly every four and a half years. Receptive bonobos are courted by many males within the group except their sons.

Young bonobos are weaned at about 4 years of age, and mature at about age 15. Mothers do the bulk of the parenting work, since it's almost impossible to tell who the father is. Like chimpanzees, infant bonobos cling to their mothers. Male bonobos usually stay with the groups they are born into, while adolescent females migrate to new groups, thus helping ensure genetic diversity.

Conclusion

If you're finding yourself feeling slightly less than unique among primates, then I've achieved my mission. If not, then it is my sincere hope that the rest of this book will convince you. I firmly believe that humans must radically recast our notions of our place in the natural order of things, if we are to survive as a species for any length of time. I also believe that acknowledging and embracing our evolved nature and rebuilding our laws and moral codes to match, might well result in the single greatest increase of quality of life that we've ever known.

Achieving this huge transformation requires nothing less than freeing us from the stranglehold of beliefs that have persisted for thousands of years and caused untold death, depredation, and destruction. Unfortunately, those beliefs are reaching a new crescendo in an age where we possess enough nuclear weapons to wipe out all life on Earth (possibly even the cockroaches) in less than an hour. I'll discuss this more in Chapters 24 through 26. Meanwhile, I urge you to keep reading with a truly open mind.

To emphasize the afterlife is to deny life. To concentrate on Heaven is to create hell. In their desperate longing to transcend the disorderliness, friction, and unpredictably that pesters life; in their desire for a fresh start in a tidy habitat, germ—free and secured by angels, religious multitudes are gambling the only life they may ever have on a dark horse in a race that has no finish line.
Tom Robbins

---

34 | *The Natural Savage*
Discovering the Human Animal

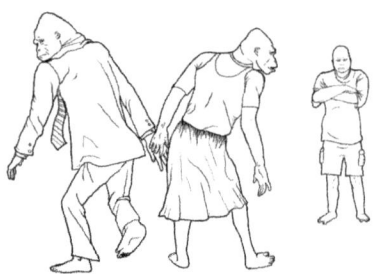

Chapter 4
All God's Creatures

> *All animals are equal but some animals are more equal than others.*
> George Orwell

> *The animals of the planet are in desperate peril. Without free animal life I believe we will lose the spiritual equivalent of oxygen.*
> Alice Walker

The many parallels between humans, chimpanzees, and bonobos are only one small tip of the giant iceberg of life. As Charles Darwin said, the separation between humans and animals is one of degree, and not of kind. All life exists on the same continuum, from the simplest single-celled organism to the most complex creatures alive today. Whether humans qualify as the most complex of all animals or not depends more on how we phrase the question than any other criteria.

Complexity aside, there is no evidence for a firm separation between humans and animals, and overwhelming evidence to the contrary. Animals treat humans like another species of animal. It is humans who base our treatment of animals on moral and ethical standards built atop an arbitrary distinction, a shining example of selfishness with no valid reasoning behind it. Those who disagree with my position point to civilization as the crowning—and uniquely human—achievement that sets us apart. To them, I reply that there is no law of nature that restricts civilization to humans, and that expanding the definition of civilization may well cause further erosion of the artificial divide between the species.

Let's very briefly explore just a few of the many ways in which humans and animals are alike, as I strive to make the case for abolishing the distinction between humans and animals.

The Stuff of Stars

A massive explosion occurred some 13.7 billion years ago. Nobody knows where the material that exploded came from or what caused it to explode. We do know that the explosion released radiation that is still detectable today as "background radiation," and as atoms and subatomic particles. Gravity caused the atoms to condense into galaxies, where stars were born, and where light elements such as hydrogen and helium formed. The explosive deaths of old stars (called *supernovas*) formed heavier elements that combined to form molecules.

Gravity eventually caused some of these molecules to come together and form planets, of which Earth is just one in a sky presumably full of countless trillions of planets. The first life on Earth evolved around 3.5 billion years ago. Nobody knows how this happened, but all life on this planet owes its existence to that first living cell.

More advanced single-celled organisms evolved and began differentiating into plant and animal species. Colonies of single-celled species gave rise to multicellular life. At some point, plant life colonized the land, followed by the first land animals that in turn gave rise to reptiles, mammals, invertebrates, and—finally—a small shrewlike creature whose descendants would become humans.

Each of us is made of the same stuff as nebulae, stars, and planets. Ponder that the next time you gaze up at the night sky.

> *There is a coherent plan in the universe, though I don't know what it's a plan for.*
> Fred Hoyle

> *In all things of nature there is something of the marvelous.*
> Aristotle

Ex Unus Plures

You already know that humans, chimpanzees, and bonobos have about 98.5% genetic commonality. What you may not know is that humans share 92% of our genes with mice and 44% with fruit flies. Experiments performed by Carol and Seymour Benzer in the 1980s found that about half of the fruit fly proteins they examined are also present in humans. Humans also share 26% of our genes with yeast, a single-celled organism. We even have 18% of our genes in common with *arabidopsis thaliana*, a plant commonly called thale cress. Many of these similarities occur because all living things have so-called housekeeping genes that keep life's many functions

> *"God" is very efficient.*
> Jennifer Cummings

humming along. They are also perfectly consistent with the idea that all life began with—and evolved from—a single cell, this planet's first living organism.

My theory of evolution is that Darwin was adopted.
Steven Wright

More evidence for the shared heritage of humans and animals can be found by examining vertebrates, animals with backbones. All vertebrates possess a skull, spine, and ribs. Skulls house brains, eyes, nose, and mouth. The spine carries and protects the spinal cord, which connects the brain with the rest of the nervous system. Ribs shield a familiar array of organs: lungs (in air-breathing animals), heart, stomach (or stomachs), intestines, liver, kidneys, pancreas, spleen, and more. Most of these organs have similar relative placements across many species.

Animals with limbs have even more features in common. Five fingers and toes are a common theme. Variations on shoulders, elbows, wrists, thighs, knees, and ankles are another. A single upper limb bone (such as the *humerus*) that connects to two lower limb bones (in this example, the *radius* and *ulna*) is another arrangement that spans a wide array of animals including humans.

The most important scientific revolutions all include, as their only common feature, the dethronement of human arrogance from one pedestal after another of previous convictions about our centrality in the cosmos.
Stephen Jay Gould

Here again, all of these similarities point to the evolution of aquatic vertebrates, some of whom eventually evolved to live on land and take to the air. Even dinosaurs have the same basic arrangement of bones and organs. Most of the species that ever evolved may be extinct, but the simple fact is that evolution picked a winning theme from the start when it came to the general layout of animal life. To paraphrase Temple Grandin in *Animals In Translation*, the process of evolution can be compared to adding new rooms onto an existing house instead of tearing it down and starting over.

It may be perfectly acceptable to think of "higher" and "lower" animals. After all, the simplest multicellular creature is far more complex than any single-celled species. It is a mistake, however, to think of any species as being more or less evolved. All life has been evolving—and continues to evolve—from the same humble beginning. Every species alive today is the product of the same 3.5 billion (3,500,000,000) years of evolution.

Intelligence

Our complex civilizations, diversity of specialization, rich languages, advanced tool usage, technology, and more all point to humans as representing the pinnacle of intellectual achievement. No animal could ever hope to earn high marks on an IQ test, college entrance exam, or vocational aptitude test. Is this proof positive of our superior intelligence? I think not.

The Triune Brain

Paul D. MacLean proposed the idea of the triune brain in the middle of the twentieth century, and further experiments conducted since then bear out his findings. Imagine having three brains in one, with each succeeding layer wrapped around the one beneath it. That is the essence of the triune brain, which consists of the following three layers:

- The *R-complex* in humans consists of the brain stem and cerebellum. This is the most ancient part of our brain, which is often called the reptile brain. This area controls our muscles, balance, and automatic functions, such as heartbeat and breathing. During periods of exertion, your R-complex signals your heart to beat faster and increase oxygen and nutrient flow to your muscles. Your higher brain can force the R-complex to cease breathing until you pass out, at which point the R-complex resumes control and you begin breathing again—a favorite scare tactic used to great advantage by some children during temper tantrums (that may also have part of their roots in this area of the brain). Some very primal emotions and instincts may also reside here. All animals with brains have this area (or similar) inside their heads; it is only common sense to surmise that they must feel some emotions.

- The *limbic system* is also known as the old mammal brain, since all mammal brains include this area. This area (which includes the amygdala, hippocampus, and hypothalamus) controls emotions, instincts, sexual behavior,

Intelligence is the wife, imagination is the mistress, memory is the servant.
Victor Hugo

I'm too shy to express my sexual needs except over the phone to people I don't know.
Garry Shandling

> *We should take care not to make the intellect our god; it has, of course, powerful muscles, but no personality.*
> Albert Einstein

and the like. It is no wonder that all humans regardless of culture or upbringing feel the same emotions and recognize them in others. The routes we take to the bedroom do vary greatly by culture, but the things we do with each other once we get there are universal. Stimulating this part of the brain produces emotions. All mammals have this part of the brain; there is no reason to think that they do not have emotions. Emotions can be fundamentally classified into pleasurable (responding to survival-enhancing stimuli) or unpleasurable (responding to survival-inhibiting stimuli).

- The *neocortex* or *cerebral cortex* is present in higher mammals and humans. This portion of the brain controls higher thinking, reason (logic), and speech. This portion of the brain appears in various stages of complexity in different species, and it is therefore appropriate to think of different animals as having differing levels and types of logical capacities.

It is important to note that each layer of the brain depends on the one beneath it. Our capacity for language means nothing without the ability to control our larynx, tongue, jaw, and lips. Our intellect is of little purpose without some guidance as to whether something is useful/good/pleasurable/survival-enhancing or not. By contrast, the higher levels of our brain can influence the lower levels, such as by temporarily suspending breathing, or by choosing to interpret a situation in a way that promotes positive or negative emotions.

> *Animals are ever so psychic. There are some people who just can't come in here. The cats particularly. They seem to know who's not right, if you know what I mean.*
> Dewitt Bodeen

In a nutshell, our most primal wants originate in the most primitive areas of our brains, while the ability to decide whether leaning over that railing for a better view is a good idea comes from more recently evolved parts of our brains.

The Frontal Lobes

Human brains have extremely well-developed frontal lobes compared to other animals. The frontal lobes are located at the top front of the brain, in roughly the area extending from your eyes up to the top of your forehead. These areas are the "gatekeepers" that direct impulses and behaviors from idea through completion. People with frontal lobe problems may exhibit impulsive behavior because of impaired impulse control. They may also have problems planning and carrying out complex tasks, and may also have emotional and/or long-term memory impairments.

Is Bigger Better?

There exists a popular myth that humans have the largest brains of any animal. This is simply not the case. The average adult human brain weighs in at about 2% of overall body weight. A 150-pound person therefore has a brain that weighs about three pounds. A sperm whale's brain tips the scales at about 17.2 pounds, an elephant brain at about 16.5 pounds. Human brains don't win any prizes for sheer weight and size.

Much is made of the so-called *encephalization quotient*, or ratio of brain weight to body weight. Here again, humanity finds itself distinctly lacking. The humble ant has a brain ratio of 50%, far larger than our own. Its 250,000 neurons are estimated to have the processing power of a Macintosh LCII computer. Having owned one of these machines, I can testify to that being an enormous amount of brain power for such a tiny creature.

Here's an interesting quirk of evolution: Neanderthals (from whom modern humans evolved) had brains that were about 10% larger than our own. Neanderthals didn't build cities or invent modern technology. Were they any less intelligent? There is a fascinating theory that may explain why our brains shrank during the most recent period of our evolution; I'll discuss this further in a later chapter.

Measuring Intelligence

Humans have neither the largest nor the most brains of all animals. What we do have is an astonishing ability for generalized and abstract thinking. It should come as no surprise that we are the highest-scoring species on tests we created to judge our own capabilities. But how well do our own tests measure true intelligence?

Howard Gardner proposed the idea of *multiple intelligences* that breaks down the generic concept of intelligence into nine specific types of learning and cognitive skills: visual-spatial, kinesthetic, musical, interpersonal, intrapersonal, linguistic, logical, naturalist, and existential. All people possess all nine intelligences in varying degrees. Someone with high visual-spatial

---

Everything you've learned in school as "obvious" becomes less and less obvious as you begin to study the universe. For example, there are no solids in the universe. There's not even a suggestion of a solid. There are no absolute continuums. There are no surfaces. There are no straight lines.
R. Buckminster Fuller

Human visual perception is a far more complex and selective process than that by which a film records.
John Berger

intelligence may learn best by seeing pictures, diagrams, and/or models, while someone with high kinesthetic intelligence may learn better in a hands-on environment.

By this definition, is the master mechanic who can diagnose, tear down, and rebuild any machine with ease any less intelligent because he received a lower score on a standardized test that disqualified him from a college education and career as a business executive? Is the business executive who can make abstract decisions based on written and verbal information any less intelligent because he his mechanical skills only go as far as filling up his car's gas tank?

Adding animals to the mix could require adding new intelligences to Gardner's list of nine. Are humans intelligent because we invented refrigeration within the last century, or are we dumb because ants and termites mastered climate control millions of years ago? Are we less intelligent because we can't, unlike a squirrel, remember where we buried thousands of nuts without a very detailed map to guide us? How intelligent are apes and monkeys that cavort and swing through the treetops, where the slightest miscalculation can spell a fatal fall? How about birds, who perform complex feats of celestial navigation with no sextant, chart, or computer to aid them?

The question of intelligence is not as easy as it first seems. By our standards, many of our planet's most successful species are dullards because they lack our diverse talents. The details are where animals excel. It is a safe bet that we would be able to recall the exact location of 30,000 buried nuts if our winter food supply depended on it. We would probably be able to determine our geographic location with an accuracy to rival today's GPS receivers if our evolution demanded annual migration. The key here is diversity versus specialization. Humans do a great many things reasonably well. Any given animal species does those few things required for its survival extremely well. Who's smarter? Depends on who's asking, why they're asking, and the context in which they're asking.

Every man feels that perception gives him an invincible belief of the existence of that which he perceives; and that this belief is not the effect of reasoning, but the immediate consequence of perception.
Thomas Reid

Perception

Temple Grandin asserts that normal people see their ideas of things, while autistic people and animals see what's really there. She cites experiments performed on airline pilots in simulators approaching ersatz runways. Having received clearance to land and performed many thousands of prior landings without incident, they failed to notice a plane on the runway blocking their path 25% of the time. The pilots expected the runway to be clear, and that's exactly what they saw. In another famous experiment, test subjects watching a televised basketball game failed to see a person in a gorilla suit running around during the show. A gorilla was the last thing they expected to see, so they didn't see it.

This difference occurs thanks to our frontal lobes, which organize perceptory data and filter it through our beliefs and expectations before sending the results of that filtering to our conscious minds. Temple Grandin refers to this as seeing *schemas* (conceptual frameworks) instead of raw sensory data. Anything that doesn't fit our expectations gets discarded as meaningless noise. This is a great thing because it helps us maintain our sense of place and order in the world, the lack of which would be extremely disconcerting. If you've ever had a powerful myth or belief shattered, then you know all too well what I'm talking about. This filtering also keeps our conscious minds from being flooded with extraneous information. Can you imagine having to hear and process all of the conversations taking place at your favorite restaurant instead of being able to focus on your dining companion?

Our filtering abilities do have their down sides, most of which stem from the rapid acceleration of technology and civilization that has outstripped our brains' ability to keep up. One is the formation and retention of negative core beliefs (see both Chapter 5 and *The Enlightened Savage*). Another is the growing possibility that our expectations are wrong. Had we evolved to fly planes, it's a safe bet that runway incursions would be non-existent. As it is, there were 31 such incidents reported in the United States during 2006.

The constant filtering of our perceptions is one of the biggest differences between humans and other animals. Animals tend to see raw sensory data, while humans filter that data based on internal criteria. This is not a hard and fast rule, but is close enough for our purposes in this book.

Nature is trying very hard to make us succeed, but nature does not depend on us. We are not the only experiment.
R. Buckminster Fuller

Emotions

My son Logan loves it when I play with him. Depending on the game, he'll either sit still with brow furrowed in deep concentration, or take off running at full tilt across the playground squealing with laughter. No one can doubt that he feels happy, excited, and content when he's receiving attention.

My partner Jennifer's cat, Tonks, frequently interrupts me as I sit up night after night working on this book and other projects. She bonks me with her head, meowing until I relent and pet her, at which point she throws herself down in an exquisitely relaxed pose. Eyes half-closed, she lolls and purrs loudly while I rub her back or scratch behind her ears. Who can deny that she feels every bit as happy and content as my son in her own way? And yet people do deny those very things.

Cherish your own emotions and never undervalue them.
Robert Henri

Chemicals known to stimulate or control emotions exist in both humans and animals, including oxytocin, dopamine, and others. (I'll expand more on this in Chapter 5.) The presence of the same substances across species that share the bulk of their genes makes it ludicrous to think that animals do not experience emotions or live emotional lives. Emotions affect and direct whatever logic a human or other animal possesses. I will even go so far as to say that logic exists for the express purpose of planning, executing, and justifying actions motivated by its emotions. Is it any wonder that advertisers target our emotions with precious little appeal to reason? We may know that the latest gizmo solves none of life's problems, but that doesn't stop us from swarming en masse to purchase it.

Charles Darwin, the father of the modern theory of evolution, did not hesitate to speculate at great length on the emotional lives of animals, a topic many scientists refuse to acknowledge to this day. Why? Is it because doing that would remove any justification for the horrors our species inflicts on others in the name of everything from barbecues to cosmetics and psychology? I forget who asked why little old ladies with blue hair and poodles know more about animal emotions than animal researchers. Reaching down to pet Tonks and seeing her obvious bliss, I know it's a brilliant question. Here again, humans have created another artificial double standard.

Let's take a very brief look at a few common emotions.

Happiness

If emotions exist to reinforce survival-enhancing behaviors, then it should be difficult to imagine that a lizard with a full belly sunning itself on a warm rock isn't experiencing bliss in its own way. Anyone who has ever lounged poolside on a warm tropical day can probably identify with that lizard.

Apes shriek, embrace, and "talk" excitedly when reuniting with long-lost friends or family members—a scene that should be absolutely familiar to anyone who's witnessed human reunions at airports or train stations. Dogs smile and wag their tails when happy or excited.

Wild dolphins perform acrobatics surpassing any aquarium show. Wild crows slide down the golden domes of the Kremlin, and monkeys on the Japanese island of Hokkaido play with snowballs. Why? Because games are fun and having fun makes one happy.

The truth is that all of us attain the greatest success and happiness possible in this life whenever we use our native capacities to their greatest extent.
Dr. Smiley Blanton

Sadness

I'll never forget the day I saw my dog Shidoni lying dead on the laundry room floor adjoining the garage and back yard, after leaving her alive and seemingly well that morning. My other two dogs are normally happy and boisterous to a fault, but they were both subdued and quiet for a long time after Shidoni's passing. I don't for a moment think it unreasonable to conclude that they were sad for their departed friend. I know I was. More than one animal has stayed with its dead young for hours or even days before moving on. Sometimes, the group or herd remains with the bereaved.

When Dr. Irene Pepperberg left her parrot Alex at the vet, he said "I love you. Come here. I'm sorry." Apes being studied at various centers have expressed sadness when companions are transferred to other facilities. Dogs whimper and cry when their owners leave them hitched outside stores and restaurants.

Pain (any pain—emotional, physical, mental) has a message. The information it has about our life can be remarkably specific, but it usually falls into one of two categories: "We would be more alive if we did more of this," and, "Life would be more lovely if we did less of that." Once we get the pain's message, and follow its advice, the pain goes away
Peter McWilliams

Aggression

My one run-in with a dog took place when I was four years old. A family friend's poodle bit me after I'd ignored repeated warnings to stop teasing her. Many species of animals display

Aggression unchallenged is aggression unleashed.
Phaedrus

aggression when competing for group status and/or mates. Some fights can cause serious injury or even death. As we saw in Chapter 3, chimpanzee tribes even wage war against each other. Animals are no strangers to aggression.

Fear

Fear is that little darkroom where negatives are developed.
Michael Pritchard

Tap the glass on an aquarium, and the fish within will start. Light a fire, and many animals will flee. Some noises such a high-pitched sounds, hissing, and clanging or banging are universally frightening to both humans and animals.

As discussed above, animals tend to filter their raw sensory data far less than humans. This makes them keenly aware of novel elements in their environments—a great way to spot a predator hiding in the bushes. This is unlike a human, who expects to see only leaves and branches, and will probably see just that until it's too late.

Novelty tends to frighten animals. One interesting side effect of this is that curiosity tends to increase with fear. Something that causes great fright (such as a cloth flapping on a usually empty clothesline), but that fails to pose a threat after a time, will be investigated and eventually accepted as part of the animal's environment. This might seem paradoxical, but actually makes perfect sense: An animal that scares easily requires some mechanism for integrating new things into its normal environment if it is to be able to function.

Love

Before I met my husband, I'd never fallen in love, though I'd stepped in it a few times.
Rita Rudner

Love is a complex emotion that involves *pheromones* (chemical signals), genotypes, imprinting, oxytocin, and dopamine. Many animals, humans included, use pheromones to attract mates. Every living thing has genes and therefore a genotype. Mammals and other animals can normally discern members of their own species. Oxytocin and dopamine are present in many species (including humans). There is therefore every reason to think that animals experience love in the romantic, filial, and fraternal senses of the term.

From the monogamy of many species to animals performing heroic deeds to rescue others of their kind or even other species, animals express love in many ways.

Grief and Despair

When Elephants Weep by Jeffrey Masson and Susan McCarthy tells many tales of animals exhibiting grief and despair after traumatic events and losses. One particularly moving passage describes elephants waking up screaming in the middle of the night—a classical sign of nightmares—after witnessing deaths and seeing tusks being cut off by poachers. Many animals of different species have been seen carrying dead young or lingering with dead companions. Is it so unreasonable to think that they are experiencing grief and despair akin to that experienced by humans when we lose someone close to us?

If we could not forget, we would never be free from grief.
Bahya Ibn Paquda

Guilt and Shame

Guilt and shame are similar, but with an important difference. Guilt refers to a specific event, such as when a dog does something it knows is wrong. The resulting display of remorse and submission can diffuse a dominant animal's anger (such as the dog's owner who discovers the misdeed) and preserves both group harmony/status quo, thus reducing the danger of harm to the wrongdoer.

Guilt is a rope that wears thin.
Ayn Rand

By contrast, shame is a more public phenomenon that requires actual or possible observation and/or judgment of others. It need not be connected with a specific event, and may in fact have more to do with an actual or perceived state of being somehow different or abnormal. Sick or injured animals may leave their herds, a seeming loss of protection that may in fact aid survival, if predators are looking for animals that stick out from the herd. In this case, a lone deer with no others around for comparison may fare better because a predator will have no basis for comparison.

Live in such a way that you would not be ashamed to sell your parrot to the town gossip.
Will Rogers

Society

Prey animals live in herds to avoid predators. Predators hunt in packs to bring down larger prey than they could alone. Some species such as bees, prairie dogs, ants, and termites band together to build hives and colonies that afford shelter, places to store and even grow food, nurseries, and even air conditioning. There indeed is strength in numbers.

Chapter 4
All God's Creatures

All animal societies have at least two key traits in common: First, all societal animals have hierarchies with dominant and subordinate individuals. Modern attempts to fit dominance theories to the way animals actually live are increasingly using terms such as respect, authority, tolerance, deference, and leadership. These terms all have emotional implications. Second, no group can function without a good deal of internal harmony. Excessive infighting would destroy group integrity and defeat the whole purpose of society.

> *Do not speak ill of society, Algie. Only people who can't get in do that.*
> Oscar Wilde

Language

As I mentioned in Chapter 3, animals don't use language in the same way humans do, but that does not mean that they lack complex means of communication that could be called language. Vervet monkeys have different calls for different kinds of predators. Prairie dogs can verbally identify specific predators (such as an individual armed rancher). Wild gorillas "talk" for a time before getting up and moving on en masse. Captive gorillas and chimpanzees have been taught sign language and a visual form of communication with nouns, verbs, and adjectives where the apes point to designated signals.

> *Language exerts hidden power, like a moon on the tides.*
> Rita Mae Brown

In what may be the most astonishing example of animal capacity for human-like language, Kanzi the bonobo has demonstrated the ability to comprehend spoken English and follow grammatical rules. *Kanzi: The Ape At The Brink Of The Human Mind* by Sue Savage-Rumbaugh and Roger Lewin contains a photograph of Dr. Rumbaugh holding Kanzi, who is gesturing in the exact direction he wants her to take him. This one image speaks volumes about animal capacity for communication.

Child Rearing

> *If your parents never had children, chances are you won't, either.*
> Dick Cavett

Many animal species raise their young, affording them protection from predators and the time to learn crucial skills. I read a story about an eagle parent teaching its young how to hunt instead of hunting on its own—a seeming paradox until one remembers that the sole purpose of life is to pass on the genes.

Ask any human parent why they raise their child, and the answer will probably involve love. If humans and animals experience similar emotions under similar circumstances, then it stands to reason that parent and young animals love each other. And why not? A strong love bond helps ensure that the young will receive the best possible upbringing.

Altruism

From human soldiers flinging themselves onto grenades to save their comrades, to a chimp who climbed up a tree to retrieve a fruit a human couldn't reach, nature is rife with altruism. Theories abound as to why this seemingly paradoxical behavior that helps one at the expense of another exists at all. Here are just a few:

It is a denial of justice not to stretch out a helping hand to the fallen; that is the common right of humanity.
Seneca

- Helping fellow members of one's family or social group helps ensure that at least a portion of one's own genes will be passed on. In other words, helping another can be a great way to indirectly win the great game of life. Animals tend to be quicker to help kin that non-relations, which lends credence to this theory.

- Altruism may help an animal curry favor and build alliances with others, which can help it gain social rank or other benefits from being associated with the "in" group.

- There are documented cases of large animals letting smaller playmates win from time to time, to keep the game going. This seemingly innate sense of fair play may also contribute to altruism. A helping animal may realize that it may need help itself someday, making "paying it forward" a good idea. It may also be spurred to help if it has received favors in the past.

- It is entirely possible that animals feel and act out of compassion because it's the right thing to do.

For whatever reason, altruism is widespread, which can only mean that nature smiles upon those who lend a helping hand.

Play

> *Of course the game is rigged. Don't let that stop you—if you don't play, you can't win.*
> Robert Heinlein

From monkeys who play with snowballs, to crows sliding off the Kremlin domes, roughhousing kittens, and my son riding his scooter as I write this, play is an important part of life. It helps develop strong bones and muscles, enhances gross and fine motor skills, and helps with hand-eye coordination. It teaches valuable social skills, and enhances dexterity and agility. It can help instill hunting techniques. Most importantly, playing is just plain fun.

Civilization?

> *Simplicity is the peak of civilization*
> Jessie Sampter

It is certainly true that animals don't build skyscrapers or freely roam the earth in planes, trains, and automobiles. We humans certainly like to think that we alone are civilized. But what exactly is civilization? The *Random House Unabridged Dictionary* offers the following definitions, to which I've added my own thoughts:

- *An advanced state of human society, in which a high level of culture, science, industry, and government has been reached, or those people or nations that have reached such a state.* Social animals have hierarchies that can be thought of as governmental. Ants, bees, and others construct marvelous homes complete with farms, waste disposal, and climate control. Some species employ specialized castes to carry out different functions.

- *Any type of culture, society, etc., of a specific place, time, or group.* Any animal that lives in groups that inhabit certain territories meets this criterion.

> *Civilization is a movement and not a condition, a voyage and not a harbor.*
> Arnold Toynbee

- *The act or process of civilizing or being civilized.* Chimpanzees and ants wage war. Given humanity's record of seeing conquering armies as bringing civilization to those they overran, this definition could arguably be applied to animals.

- *Refinement of thought and cultural appreciation.* We can't know for certain what goes on in animal minds, except by observing their actions and monitoring their brain activity. We like to think that our own thoughts are refined and that we appreciate culture; in short, we revel in our humanity. It is certainly possible that animals experience joy in their own natures, in their own fashion. Our inability to read animal thoughts and judge their sense of culture does not disprove the existence of either refined thought or cultural appreciation in animals. And what of the bowerbird, whose reproductive success depends on aesthetics?

- *Cities or populated areas in general, as opposed to unpopulated or wilderness areas.* Animals that live in groups or within certain territories certainly fit this criterion. If humans define a wilderness as an area devoid of human population, then a human city may just qualify as a wilderness to an animal.

- *Modern comforts and conveniences, as made possible by science and technology.* Here I may have to admit defeat, for animals have no discernible science or technology. Except, of course, that their dwellings are marvels of form and function. I doubt any animal looks to humans for engineering ideas, but we sure look to them.

Are animals civilized? Remove the anthropocentric overtones from the definition of civilization, and my answer is a resounding maybe.

Nature is just enough; but men and women must comprehend and accept her suggestions.
Antoinette Brown Blackwell

Conclusion

Whether or not this brief exploration causes you to reevaluate your own ideas about humans and animals is in many ways irrelevant. Your highly developed frontal lobes may be filtering more of your reality, and thus buffering you from your own inner animal nature, but that does not make you any less of an animal. From genetics to your behaviors and emotions, your belonging to *homo sapiens* does not make you unique.

As I said in Chapter 3, I believe that the overwhelming majority of human problems and suffering stem from our unwillingness to know and embrace our inner animals, our inner savages. Understanding that the human animal is an animal, learning how that animal operates, and actively modeling our societies around this knowledge may well be the key to transforming all of our lives for the better. That can't happen too quickly—for all of our sake.

I believe in God, only I spell it Nature.
Frank Lloyd Wright

Chapter 5

What is Reality?

> *"Reality" is the only word in the English language that should always be used in quotes.*
> Unknown

It is tempting to think that we are directly connected to—and perceive—the world around us exactly as it is. We therefore want to believe that everything we see, hear, feel, smell, and taste is real, exactly as we perceive it. Our reactions to people whose notions of reality differ from our own range from agreeing to disagree to locking them in secure psychiatric facilities. The old adage that 20 people witnessing the same car crash will provide 20 different descriptions is more accurate than we sometimes want to acknowledge. As I have said before, there is no such thing as objective reality. Everything we know and perceive about our surroundings arrives in our conscious minds pre-filtered by our beliefs.

Perception at a Distance

> *All perception of truth is the detection of an analogy.*
> Henry David Thoreau

Your body is equipped with a vast array of sensors that detect light (sight), sound (hearing), physical sensations (touch), and chemicals (smell and taste). All of this raw information proceeds to the brain as a series of impulses traveling along a network of nerves that eventually reach the brain. Here, this information is sorted by type and location, and either forwarded for conscious processing, or suppressed. For example, you are aware of a chair pressing against you when you first sit down, but may cease to be aware of that sensation as you go about whatever it is you are doing. The chair is still pressing against you, but you have ceased to be aware of the information the nerves in your backside are continuing to send up.

Everything you are aware of seeing, hearing, feeling, smelling, and tasting has been processed and interpreted long before you even became aware of what you are sensing. The pain you feel when you stub your toe is all in your head; your brain interprets the signals it receives as pain in your toe, and you feel your toe hurting. This is an illusion, because your brain is where the sensory data from the nerve endings in your toe are perceived as pain. People with some types of nerve damage (such as spinal cord injuries) have no sensation below the point of injury. Your brain is what feels, not the point where the sensation occurs.

It is the mind which creates the world about us, and even though we stand side by side in the same meadow, my eyes will never see what is beheld by yours, my heart will never stir to the emotions with which yours is touched.
George Gissing

In short, everything you are aware of about the outside world and your interactions within it, is nothing more or less than what you believe it to be, in accordance with your core beliefs that form your most fundamental truths. So how does raw sensory input get converted into a person's reality?

A Living Computer

Every thought, feeling, belief, and memory is completely dependent on your brain's capabilities and limitations. Your entire life and individual reality depend on what your brain can and cannot do.

A ready-to-use computer consists of both hardware and software. Computer hardware is the machinery itself. The software is simply electromagnetic information stored and processed by the hardware. Software must have functioning hardware in order to run. It also depends on the hardware for its capabilities. If your computer lacks speakers, then you won't hear anything from the audio program you just installed, no matter how much you adjust the program's settings.

One need not be a chamber to be haunted. One need not be a house. The brain has corridors surpassing material place.
Emily Dickinson

A human brain is roughly analogous to computer hardware that stores and executes the many software programs that combine to form all of your thoughts, feelings, beliefs, and memories. If we are nothing but biological creatures whose brains are the source of our consciousness, then we are constrained by our brain's abilities and limitations. If we possess a

If the brain were so simple we could understand it, we would be so simple we couldn't.
Lyall Watson

soul or spirit, and our brains are merely conduits for consciousness, then we remain constrained by the same limits, in the same way that one cannot obtain a high-definition picture from a 1950s TV set. In either case, comparing a brain to a computer can be very helpful, provided that we remember to draw only loose functional parallels from said comparison.

If a man is offered a fact which goes against his instincts, he will scrutinize it closely, and unless the evidence is overwhelming, he will refuse to believe it.
Bertrand Russell

The term *hard-wiring* refers to building a specific program into a computer's actual physical circuits. I believe that our prey instincts and images are hard-wired into our brain's circuits on a physical, genetic level. Chapter 4 presented several examples of behaviors that are controlled by genes. Chapters 6 through 9 will explore how prey animal habits revolve around avoiding predators, and why that is so important for understanding human nature. For now, let's focus on what motivational speakers call the *process of realization*, the process by which you build your unique personal reality from your sensory input.

The Process of Realization

Every man's work shall be made manifest.
1 Corinthians

The process of *realizing* (building, or making real) your personal reality from raw sensory input follows the acronym ETEAR as follows:

- **E**ARLY EMOTIONS caused by one's earliest experiences lead to
- **T**HOUGHTS that form our deepest beliefs about the world and our place in it, which in turn fuel a cycle of addictive
- **E**MOTIONS that have the same effects on our bodies as addictive drugs and therefore drive our
- **A**CTIONS that lead to the consequences that form our
- **R**EALITY, which reinforces our THOUGHTS and begins the cycle all over again at that point.

Let's explore this process in more detail.

Early Emotions

Emotions have taught mankind to reason.
Vauvenargues

Humans are emotional beings. Only 5-6% of the 2,000 or so bits of information we are conscious of each second are logical; the rest are emotional. We don't think our way through life nearly as much as feel our way through. It is certainly plausible that our emotional lives begin before we are born.

At their most fundamental level, emotions consist of chemical compounds. These compounds travel throughout our bodies in our bloodstream, since it is impossible to contain blood-borne substances to any specific location in the body. These compounds attach to opiate receptors in our cells—the same opiate receptors that respond to addictive drugs, such as heroin. This has two clear implications: One, that we experience our emotions throughout our bodies. Two, that we become physically addicted to certain emotions and emotional patterns.

A torn jacket is soon mended; but hard words bruise the heart of a child.
Henry Wadsworth Longfellow

Human embryos and fetuses create both the chemical compounds and opiate receptors as they develop, hence the idea that they may be developing addictions to certain emotions before birth. Certain stressors, such as loud and/or high-pitched noises or the mother's rapid heartbeat—and possibly even her own emotional chemicals—may well be giving the developing human its first impressions of the world through a complex interplay between raw sensory input, genes that influence how that input is interpreted, and chemical compounds. This emotional conditioning (what we'll call programming in a later chapter) continues after birth, rapidly becoming emotional addiction—the physical craving for the specific chemical compound(s) that the baby is addicted to just like any junkie is addicted to illicit drugs.

Children might or might not be a blessing, but to create them and then fail them was surely damnation.
Lois McMaster Bujold

Thoughts

As the child grows, emotional conditioning drives the creation of core beliefs, deeply held thoughts about the nature of the world and normality. These thoughts are logical conclusions on a deep, usually subconscious level. By logical, I mean on the order of "if A, then B, else C," and not on the order of a rationally constructed hypothesis assembled from available evidence. Psychologists often refer to logic like this as *magical thinking*.

You cannot educate a man wholly out of superstitious fears which were implanted in his imagination, no matter how utterly his reason may reject them.
Oliver Wendell Holmes Jr.

These core beliefs define truisms that hold whatever types of circumstances best feed the emotional addiction mentioned above as normal and even optimal.

Emotions

> *Where we have strong emotions, we're liable to fool ourselves.*
> Carl Sagan

Back to emotions we go. Circumstances that don't produce the emotional chemicals we are accustomed to, fail to satisfy our emotional addictions, and are therefore deemed abnormal. It is important to note that most—if not all—of this occurs on a deep subconscious level, meaning that we are all too often unaware of our internal patterns. Like any good junkie, any person who goes too long without their "fix" will start doing all they can to rectify the increasingly dire situation.

Actions

> *We always do what we most want to do, whether or not we like what we are doing at each instant of our lives. Wanting and liking many times are not the same thing. Many people have done what they say they didn't want to do at a particular moment. And that may be true until one looks deeper into the motivation behind the doing.*
> Sidney Madwed

One may consciously enjoy circumstances outside one's norm. For example, someone raised in an abusive relationship may derive tremendous joy from a loving and supportive relationship. The problem is that the person has long since become addicted to pain, anger, and other negative emotions, and will start missing those chemical compounds. Eventually the withdrawal becomes so strong that the person finds a way to restore order, in the same way that an addict will find any way possible to pay for that next dose.

Returning to the addiction-prescribed norm may not be pleasant. In fact, it may be extremely painful, and contribute to loss of self esteem and other deleterious effects. Few junkies get true pleasure from their addictions beyond those first euphoric highs. Still, like moths to the flame, they keep coming back again and again, because they are powerless to combat the addiction.

> *I've arrived at this outermost edge of my life by my own actions. Where I am is thoroughly unacceptable. Therefore, I must stop doing what I've been doing.*
> Alice Koller

Addictive substances have another insidious property: Over time, the body builds up a tolerance or resistance to the substance, meaning that ever-higher doses are required to achieve the same effect. Look at your own life. I'll bet that you see the same recurring themes or patterns in your own past and that life's ups and downs have tended to become more pronounced over the years. I'll go one step further, and postulate that these swings have occurred with almost alarming regularity. If you're not seeing this pattern occurring in your own life, then either you've done something to alter it, or you're not looking hard enough. Don't feel too bad if you can't identify such a pattern; it can manifest itself in many different ways, to the point where it's no longer obvious to you.

Results

Every action, no matter how small, has a consequence. Your actions—or in some cases inactions—have created consequences that you are living with today. Your brain judges these consequences against your core beliefs. Consequences inconsistent with core beliefs are deemed unacceptable, no matter how wonderful they may seem to the conscious mind. Consequences that fit your core beliefs are deemed good. Your core beliefs are validated, which fuels your emotional addictions, and the process starts over.

In nature there are neither rewards nor punishments. There are consequences.
Robert Ingersoll

Which foods are good to eat, who might make an ideal mate, politics, religion, philosophy, current events, history, moral values, ethical judgments, and all of the thousands of preferences and biases that find favor or disfavor with you are the end result of the process of building your own personal reality. If you believe that the Universe and all life as we know it sprang into being over the course of a week a few thousand years ago, then you're probably not reading this book. If you are, chances are excellent that you find me less than credible as an author.

The happiness of most people we know is not ruined by great catastrophes or fatal errors, but by the repetition of slowly destructive little things.
Ernest Dimnet

Reality Defined

The WordNet database operated by the Cognitive Science Laboratory at Princeton includes, "all of your experiences that determine how things appear to you" among the definitions of *reality*. To expand on this definition, your personal reality consists of the combined consequences of your actions (which can include both physical action or inaction, and conscious thoughts or feelings). Your actions are driven by your need to satisfy your emotional addictions that stem from—and lend credence to—your core beliefs, which in turn arose from early emotional programming. This is your subjective personal reality, and the subjective personal reality of all people.

Reality is merely an illusion, albeit a very persistent one.
Albert Einstein

There is also an objective shared reality, which are those things perceived by most people. For example, most people holding this book would agree that they are holding a book and can probably read the writing and grasp the points being made. How each reader reacts to the content in this book is a perfect example of subjective personal reality.

Opinion is ultimately determined by the feelings, and not by the intellect.
Herbert Spencer

PART TWO

Predator Avoidance

58 | *The Natural Savage*
Discovering the Human Animal

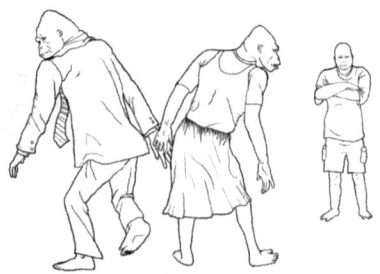

Chapter 6

Rise of the Prey

There are many species of monkeys and apes. Some live in trees, while others live in grasslands and in various other climates. A close examination of each species reveals that each is perfectly adapted for its unique habitat. Japanese snow monkeys can even swim—very useful for island living. Humans evolved from apes who in turn evolved from a species that gave rise to monkeys and apes in both the new and old worlds. Let's take a look at our evolutionary history and how our prey ancestry drives everything we do.

The most important scientific revolutions all include, as their only common feature, the dethronement of human arrogance from one pedestal after another of previous convictions about our centrality in the cosmos.
Stephen Jay Gould

Our Evolutionary Tree

The story of human evolution begins about 35 million years ago, when the line that would become New World monkeys such as *cebinae* (capuchin monkeys), *actinae* (night monkeys), *atelinae* (howler monkeys), and *pithecinae* (uakaris and sakis) split from the line that would eventually lead to Old World monkeys, apes, and humans. Other relatives, such as tarsiers, lemurs, and lorises had already branched off this line.

The next major split in our evolutionary history occurred between 13 and 16 million years ago. One side of this split led to orangutans (*pongo pygmaeus*), gorillas (*gorilla gorilla*), and other

Old World monkey and ape species, while the other continued toward the next major branch of our evolutionary tree.

The next major split occurred some 5 million years ago, when the line that would become humans split from the line leading to chimpanzees (*pan troglodytes*) and bonobos (*pan paniscus*).

The human line evolved two primitive hominids (*australopithecus africanus* and *australopithecus robustus*) that flourished on the African savanna from roughly 4 million to 1.2 million years ago. The *robustus* line

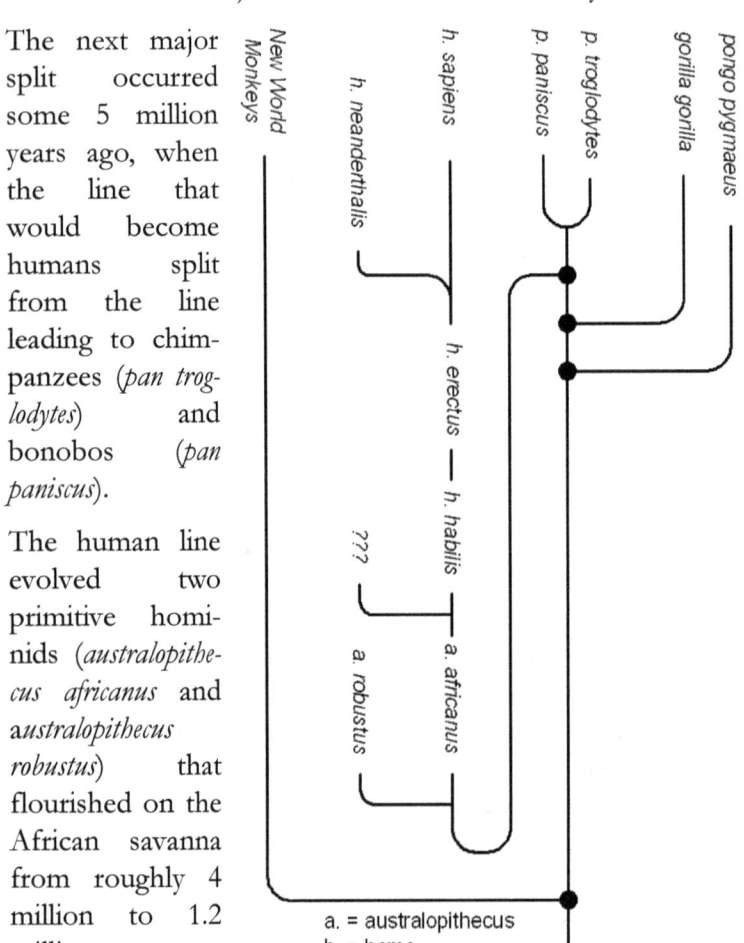

a. = australopithecus
h. = homo
p. = pan

It's evolve or die, really, you have to evolve, you have to move on otherwise it just becomes stagnant.
Craig Charles

died out, leaving the other line to evolve into *homo habilis* and a possible other hominid species that is as yet unnamed. *Homo habilis* lived from about 2.6 million to 1.4 million years ago. *Homo erectus* appeared about 1.8 million years ago, and survived until only about 100,000 years ago.

Neanderthals (*homo neanderthalis*) lived from about 150,000 to about 35,000 years ago. Their line split from a primitive form of *homo sapiens* that existed from about 300,000 to 120,000 years ago. Another split led to modern Asians. Meanwhile, the main evolutionary line led to modern Africans and Cro-Magnon Man, the ancestor of the modern Caucasian race.

These are just the highlights of our evolutionary history. I have left out many other species (such as *homo heidelbergensis* and *homo ergaster*) that played a role in the long march toward what we know of today as humanity.

The chart does not show the split between Asians and the line of Africans and eventual Caucasians. Each of these subsets of modern humans is adapted for a different climate. For example, dark African skin may be better for warding off sunburn in the tropical latitudes. Narrow Asian eyes may reflect adaptations for life in Mongolia, where winds fling grains of sand and dirt. The pale sun of higher northern latitudes, plus the need for clothing, may account for pale Caucasian skin.

I can only speculate that these racial adaptations may have represented the beginnings of three separate species of humans, just as accumulated adaptations led to the differentiation of common (*troglodytes*) and pygmy (*paniscus*) chimpanzees. I say "may have" because modern technology is enabling global travel and providing millions of opportunities to interbreed that could well prevent this from occurring. If I'm right, then this marks the first time in history that a species has forestalled its own fragmentation and diversification.

On this point, I must be perfectly clear: While I see the beginnings of new species in the various races of humans, I do so without any judgment or prejudice about any supposed superiority of one race over another. On the contrary, the shared heritage of all modern human races proves their equality without the need to invoke any moral or ethical values. Even if that were not true, common decency demands that we recognize our fellow humans as no more or less than ourselves.

Lastly, this chart and accompanying history do not include the recent discovery of *chororapithecus abyssinicus* that could move the split between humans and apes back several million years from the dates listed. I will update this book should further discoveries make that necessary.

Like leaves on trees the race of man is found, now green in youth, now withering on the ground; another race the following spring supplies: They fall successive, and successive rise.
The Iliad

The condition upon which God hath given liberty to man is eternal vigilance.
John Philpot Curran

Humans as Prey

I have already mentioned the jaw mutation that caused human jaws to weaken some 2.4 million years ago. Animals with weak jaws lose both the ability to bite and an important defense against predators. Monkeys and apes are far stronger than humans on a pound-for-pound basis, while most humans can outwit any monkey. In essence, humans swapped brawn for brains. How well did this work? Look around you. Chapter 5 summarized some of the portions of *The Enlightened Savage* that explain how our ancient survival instincts can keep us from attaining the lives we want and deserve. But I digress.

The bottom line is that humans are prey animals. Humans and monkeys evolved from the same common ancestor. We share the same evolutionary roots. All monkeys and apes are subject to predation, as we saw in Chapter 4.

It is certainly true that man and apes can be predators. Chimpanzees hunt bush babies using crude spears, and we commonly refer to primitive human tribes as hunter-gatherers. Still, the threat of death by tiger or snake only recently ceased to be a daily threat for the majority of people alive today. An unfortunate few are still killed and eaten every year, and our ape cousins face possible predation at any moment of any day. Modern civilization has only existed for a few thousand years, while our prey ancestry stretches back for at least a few million years. Our prey instincts are therefore alive and well, operating in the manner described in Chapter 5.

Monkey Predators

Monkeys and apes face predation from big cats (lions, tigers, leopards), snakes, and even birds. Panama's harpy eagle is both the world's biggest bird of prey and one of the few built to fly through the forest canopy carrying heavy loads, instead of over it. Put yourself in a Central American monkey's position. One second you're perched on a branch enjoying a piece of ripe fruit, the next you're caught in piercing talons that cut into you and prevent your escape until the bird decides to eat you. I'm not making this up; harpy eagles feed on monkeys.

Monkey predators live on the ground, in the trees, and in the air. There is no place for a monkey to run to or hide where it can be truly safe. This is the human evolutionary legacy.

Prometheus Rising

Humans learned to harness fire about 500,000 years ago, arguably the single greatest technological leap of all time, since fire

We are just an advanced breed of monkeys on a minor planet of a very average star. But we can understand the Universe. That makes us something very special.
Stephen Hawking

I want to find a voracious, small-minded predator and name it after the IRS.
Robert Bakker

Labor to keep alive in your breast that little spark of celestial fire called conscience.
George Washington

Chapter 6
Rise of the Prey   63

in one form or another is either directly or indirectly responsible for just about everything we take for granted today.

Combine a snake's lithe body and scaly skin with a big cat's legs and fangs, a raptor's wings, and fire and you get a mythical creature that appears in virtually all past and current cultures on Earth: a dragon. The existence of dragons and dragon-like creatures is a worldwide phenomenon, this among people who routinely kill each other over both differing beliefs and flavors of beliefs (such as Catholics and Protestants).

Humans recognize cats, snakes, and eagles very easily and are prone to fear them. I believe this is because they are our evolutionary predators. We also experience powerful emotions around fire, emotions, which doubtless began the first time one of our ancestors beheld a blaze and continue to this day. If you've ever sat around a campfire, fought fire, or been anywhere around fire, then you know exactly what I mean. Homes with fireplaces fetch more than those without them at sale time, all else being equal.

It stands to reason that the mythical dragon slayer may be the origin of today's tales of superheroes. And why not?

Primeval Evidence

Holes in a two-million-year old *australopithecus africanus* skull known as the Taung child skull are consistent with being killed by a predatory bird, such as a large eagle. Paleoanthropologist Lee Berger studied the Taung child, saying that such discoveries are, "key to understanding why we humans today view the world the way we do." I must agree.

Predator/Prey Behavior

Place a prey animal into a new situation, and chances are it will move hesitantly and choose to retreat or cower if threatened. A predator will probably venture forth both sooner and more boldly than a prey animal. Startle a prey animal, and it will shrink back. Startle a predator, and it may pounce, as anyone who has ever played with a kitten can attest.

Prey animals are always vigilant, always on guard, always taking steps to avoid detection, track their potential killers, or

---

The inner fire is the most important thing mankind possesses.
Edith Sodergran

We often give our enemies the means of our own destruction.
Aesop

The optimist proclaims that we live in the best of all possible worlds; and the pessimist fears this is true.
James Branch Cabell

both. They tend to be nervous and on edge, ready to take flight at a moment's notice. Social prey animals often employ one or more lookouts. The meerkats at the San Francisco Zoo are a prime example: The odds of one of them falling prey to any other animal are virtually nil, but that doesn't deter them from always stationing a lookout atop the tallest point in the enclosure. I happened to be there one day when the lookout climbed down from his viewpoint. He was replaced within seconds. Evolution favors those prey animals that keep careful watch.

The sun, the moon and the stars would have disappeared long ago had they happened to be within the reach of predatory human hands.
Havelock Ellis

Predators are altogether different. Lions and tigers relax and sleep in plain view of anyone. Bald eagles rest on bare branches, also in plain view. After all, no other animal is going to take them on (except humans, and that only very recently in evolutionary history, thanks to vastly superior technology). In this case, evolution favors those animals who can relax between hunts to digest their last meals and gather strength for the hunt to come.

Like most rules, this one is rife with exceptions. Animals without natural predators are often nonchalant to the point of being almost oblivious, while their counterparts around the world live in mortal fear. Examples include penguins who allow humans to approach on land, and the dodo bird whose placid nature led to its downfall. This too makes evolutionary sense: If there is nothing to fear, then why be afraid? Fear consumes energy that could be spent reproducing or foraging. As for the now-extinct dodo bird, there is no way that evolution could have foreseen boatloads of humans disembarking to wreak wholesale dodocide.

Fear in Action

Fear is a good thing in moderation. Healthy fear helps us predict a possible future (if I walk in front of that lion it may kill and eat me), and inhibits actions that could bring the object of that fear to pass. I believe that all prey animals, humans included, are predisposed to fear and the spectrum of negative emotions that stem from fear. Our vocabulary contains many words to describe negative emotions, but relatively few to describe positive ones.

Fear is that little darkroom where negatives are developed.
Michael Pritchard

We can see evidence for the humans-as-prey argument by looking at what we fear. Plenty of people fear snakes, even though the risk of dying from snakebite is extremely small for most of us alive today. The Centers for Disease Control reported 42,443 deaths from traffic accidents in the United States in 2001. That's 4 per hour. Number of people killed by snakebite? 12. That's 1 per month. In other words, you are about 2,880 times as likely to die in a traffic accident as you are from a snakebite.

I know this, and yet I hardly ever give a second thought to hopping in a car. I do, however, tend to avail myself of every opportunity to avoid handling a snake, even a harmless garter snake. Tonks the cat routinely pounces on and bites me in play. Doesn't bother me in the slightest. The idea that a snake I'm holding could bite me? No thanks. Don't laugh! Think about this a moment, and you'll soon realize that your own fears are just as senseless when held before the cold light of reason. Why? Because we evolved with snakes. Compress human evolution into a single hour, and cars have only been with us for about 1/10 of a second. Snakes have been with us the full hour.

On the other hand, the average person comes near many thousands of cars in a lifetime, while only encountering a few wild snakes. Snakes may therefore pose a much higher "per capita" risk than cars, meaning that the chances of dying from one snake encounter may be much higher than the chances of dying from one encounter with a car (such as crossing a street). I have not crunched any numbers to examine this counterpoint, but I cannot ignore the possibility that it represents the more accurate scenario and explanation for why people fear snakes so much more than cars.

Fight or Flight?

Isn't it interesting that only a single letter separates these two opposing words? Even more interesting, humans have two types of fear. The first, fast fear, responds to stimulus within 12 milliseconds. The second, conscious fear, responds within about 24 milliseconds to let us know what we're scared of. A scared human is in motion before s/he even knows that s/he is afraid, let alone what s/he is afraid of.

Man must get his thoughts, words and actions out of this vast moral jungle. We are not predators. We are, hopefully, more than instinctive killers and selfish brutes. Why take such a dim view of our potentialities and capabilities?
H. Jay Dinsah

Part of the happiness of life consists not in fighting battles, but in avoiding them. A masterly retreat is in itself a victory.
Norman Vincent Peale

As we discussed in Chapters 4 and 5, humans consciously perceive ideas about what's going on around them, ideas that are in turn at least partially based on each individual's core beliefs. Animals presumably lack this filtering, and thus perceive and take the world far more literally. Interestingly enough, this leads to a seeming paradox wherein increased fear also tends to lead to increased inquisitiveness. The more frightened the animal, the more curious it will also be.

The explanation is actually fairly simple: Fear consumes energy and time. An animal paralyzed by fear runs the risk of malnutrition and stress-induced illnesses. Therefore, if the initial fear proves unfounded, the animal will experience a growing desire to investigate the object of its fear, a desire that will be acted upon once the curiosity becomes stronger than the fear. The animal will eventually realize that said object is harmless, incorporate that knowledge into its sense of normalcy, and cease being scared of it. This allows it to maintain a high level of vigilance without losing the ability to function.

The Bottom Line

Humans evolved from prey animals and remain prey animals at heart. Individuals of all ages are fair game for predators. It is therefore impossible to switch the prey response on and off, as can be done for certain types of learning, such as walking and talking, which we'll discuss in Chapter 8. We live like the monkeys we evolved alongside for millions of years, and our modern adaptations are designed for that bygone era.

Predators, small roving bands, scarce food, and group dynamics and politics are the environment we evolved in until extremely recently in our history. Our evolved instincts are therefore alive and well, and still driving everything we do.

We are responsible for what we are, and whatever we wish ourselves to be, we have the power to make ourselves. If what we are now has been the result of our own past actions, it certainly follows that whatever we wish to be in future can be produced by our present actions; so we have to know how to act.
Swami Vivekananda

Chapter 7

Predation 101

> *God created a number of possibilities in case some of his prototypes failed—that is the meaning of evolution.*
> Graham Greene

From an evolutionary standpoint, you are alive for one reason and one reason only: to pass your genes on to the next generation, either directly (through parenting) or indirectly (such as helping care for a relative). You yourself are nothing but the means, the tool by which your particular sequence of deoxyribonucleic acid (DNA) either survives or perishes. Those individuals whose genes survive them have won the great game of evolution, while those who die without having reproduced are the losers. Don't believe me? Consider for just a moment that you owe your very existence to an unbroken string of successful reproduction that extends all the way back to the first living organism on Earth. One break in this chain, and you would not be reading this, because you could never have been born.

Considering that you evolved from (and remain) prey, it seems reasonable to assume that at least a few of your ancestors survived encounters with predators. I find it more than a little humbling that I probably owe my life to an *australopithecus* who barely managed to leap into a tree just as a cat was about to strike. It is utterly impossible to reproduce once you've been reduced to chewed lumps of flesh inside a predator's belly. I therefore assert that predator avoidance is the number one priority for any species of prey animal, humans included.

Think about it: You can probably put off eating, drinking, sleeping, leaving your home, having sex, or just about anything you may have going on in life. Nobody has the luxury of procrastinating when faced with a predator. This is precisely why your body's "fast fear" mechanism exists: to start getting you out of harm's way before you can waste time asking questions.

Let's look at the predation process and some of the methods prey animals have to avoid becoming lunch.

The Predation Process

There is no such thing as a free lunch. Any hungry predator must follow a long process that may or may not conclude with a full belly. This predator must therefore weigh the costs of attempting to catch lunch against the odds of investing that effort on a case-by-case basis. Every failed effort reduces the amount of resources available for subsequent attempts. The predator must win the evolutionary game of passing on its genes at the expense of other animals, who are doing all they can to win their own games. Prey animals don't want to be killed and eaten, and they will do all in their power to avoid that fate.

My centre is giving way, my right is retreating, situation excellent, I am attacking.
Ferdinand Foch

Every hungry predator must successfully execute the following process in order to end up with a meal:

1. *Searching.* Hunger (empty stomach) triggers an emotional response that motivates the search for prey. Should the predator go out looking for its next meal, or lie in wait for that meal to wander past within striking distance?

2. *Detection.* Does the predator rely on sight, hearing, smell, touch, or some combination of senses to sense the presence of possible prey?

3. *Acquisition.* Merely knowing that a potential prey animal is in the area isn't good enough. Where exactly is this potential prey?

4. *Firing Solution.* In military terms, a *firing solution* is a set of instructions that direct a weapon to hit its intended target. For example, a naval gunner must know the target's distance, direction, bearing, and speed in addition to the weather, effect of the Earth's rotation (Coriolis effect), etc.

in order to properly aim his guns. A predator must determine how to approach its prey, which individual in a group will be attacked, and more. A frog must know a passing fly's distance, direction, and speed of flight in order to catch it with its sticky tongue. Every predator must therefore create its own version of a firing solution based on the unique circumstances of each potential attack.

5. *Attack*. The predator must launch its attack at some point, and maintain it until the prey is brought down, or the effort ceases to be worth the potential reward.

6. *Kill*. Catching a prey animal does not guarantee success. The prey may fight, squirm, or otherwise avoid being killed (such as if its companions come to its aid). It may even inflict damage on its attacker. For example, one well-placed kick from a zebra is enough to ruin any lion's day.

7. *Defense*. Making the kill still isn't enough to guarantee the predator a meal. Individuals of its own or other species may try to steal some or all of its hard-won catch.

One break in the preceding chain of events means that the predator fails to catch or eat its prey. That prey therefore has every incentive to make the first six steps as difficult as possible for all would-be predators. Unfortunately, the prey animal has no control over the seventh step, because it's already dead by then. Prey animals do, however, have plenty of tricks up their sleeves when it comes to defeating some or all of the first six steps of the predation chain.

Searching

Predation is driven by hunger. Animals that kill indiscriminately put their food supply at risk. Therefore, with very rare exception, animals do not hunt for sport. Hunger spawns emotions that in turn motivate the predator to begin hunting. Absent this hunger and associated emotional motivation, the predator will not hunt, and prey animals are perfectly safe.

All prey animals need to know their chief predators and the methods those predators use to hunt them. First of all, which animals are predators, and what distinguishes them from other, harmless animals? Does the predator go looking for its prey or wait for its lunch to come to it? If the predator hunts,

The most violent appetites in all creatures are lust and hunger; the first is a perpetual call upon them to propagate their kind, the latter to preserve themselves.
Joseph Addison

does it swim, slither, walk, or fly? If it lies in wait, what are its most common hiding places?

These are crucial questions. It does an animal little good to scan the ground around it if its main predator is an eagle. Likewise, a fish whose predators prefer to hide may do well to avoid swimming too close to rocks and corals that can provide easy concealment.

Does the predator behave differently when hunting versus merely being out and about? If so, what are the differences in posture, movement, facial expressions, calls/vocalizations, scent, etc? Being able to distinguish a predator that's hunting from one that isn't can save a prey animal needless stress.

Knowing what one's predators are and how those predators hunt can go a long way toward minimizing the risk of being killed and eaten. Human ancestors had a particularly tough time, because their predators hunted on the ground, in the trees, and in the air. Some cats hide in bushes, others approach slowly before launching a final sprint, and still others wait in tree branches for unsuspecting meals to wander past underneath. Some eagles can hunt under the forest canopy, while others prefer to fly over open terrain. Snakes may lie in bushes, under rocks, or in tree branches. Our ancestors had few safe havens.

Detection

How does a predator detect its prey? Does it rely on sight, hearing, smell, feel, radiated heat, electrical discharges, or other means?

Camouflage can conceal an animal from visual detection. A famous example in the British Isles concerns *biston betularia* (commonly called the peppered moth). This moth is normally light-colored, perfect for blending into pale tree trunks. Pollution darkened many tree trunks, rendering the pale moths easily visible, while allowing darker-colored individuals to escape unnoticed. Other insects (such as *carausius morosus*) resemble sticks, while thorn bugs (*umbonia crassicornis*) convincingly resemble thorns. Many animals of all types are colored to blend into their natural environments to make detection harder.

Look and you will find it. What is unsought will go undetected.
Sophocles

If you can't avoid detection, why not try convincing potential predators that you're poisonous or otherwise unpalatable? The false cobra (*malpolon moilensis*) does this by mimicking the more venomous king cobra, while the tasty viceroy butterfly (*limenitis archippus*) mimics the inedible monarch butterfly.

If the predator relies on hearing, then the prey animal would do well to remain as silent as possible. This may entail following established paths that tend to be clearer of leaves and other debris than undisturbed ground, moving very slowly, or rarely moving at all. Prey whose hunters rely on smell may mask their scent by burying their feces, coating themselves with false scents, or emitting foul smells.

Let's return to sight for a moment. Does the predator respond to its prey's shape or to motion? Some birds of prey may see through a potential target's camouflage to see its actual shape and attack the shape, while others may detect motion and dive before positively identifying the target, if the motion is consistent with a prey animal. Staying under cover until one is reasonably sure that the skies are clear before making a very short dash may be a viable strategy in this latter case.

> *He can run but he can't hide.*
> Joe E. Louis

Experiments with human armies revealed that color-blind soldiers can more easily detect camouflaged enemies than those with normal color vision. It is possible that the adaptations that allow us to spot ripe fruit detracted from our ability to spot hidden opportunities or threats. This is a perfect example of evolutionary trade-offs at work.

Humans have faced some rather unique problems during our evolutionary history: We are not particularly good climbers, nor are we blessed with any form of natural camouflage. Many predators possess far superior sight, smell, and hearing and can easily outrun, outmaneuver, and outfight us. It's hard for humans to escape detection.

Acquisition

> *The mice which helplessly find themselves between the cats teeth acquire no merit from their enforced sacrifice.*
> Mahatma Gandhi

A predator that has detected possible prey in its vicinity must now ascertain that prey's exact location. A predator that sees its prey may well accomplish both detection and acquisition at once, while one that has caught a scent must determine where that scent is coming from and how fresh it is. Prey must also be aware of the signals they may be giving off, everything from standing out against the backdrop of their environment to the smells they are leaving behind.

A true-crime show I watched one evening told the story of a bloodhound tracking a fugitive. This by itself was nothing new. What shocked me is that the bloodhound was able to follow the fugitive's path for many miles along a freeway. Everyone sheds skin cells, and the bloodhound was able to follow the scent of skin cells that wafted onto the freeway.

Prey animals may be able to smell predators and determine how far away danger lies, or how long ago the scent was laid. There is little need to panic over a distant or old threat, because there is no immediate danger. The prey can keep its distance and possibly avoid detection. Other prey animals have lookouts that scan the environment and alert the group. The individual lookout may be at increased risk for predation, but the group as a whole is much safer. The lookout may even help pass along some of its genes by being killed and eaten, if doing so ensures that some or all of its kin survive. Remember, evolution doesn't care one bit about the individual.

Humans lack the visual acuity of many predator animals, and our sense of smell also lags way behind. A person's chances of knowing a predator's exact location before it figures out their exact position is therefore remote. Add in the fact that we perceive what we believe we perceive (see Chapter 4), and it becomes obvious that humans stand little chance of not being hunted, detected, and acquired by potential predators.

Firing Solution

Exactly how and when should the predator attack its prey? Is it hunting alone or with others? If flying, should it maneuver to dive out of the sun, thereby blinding the prey, or should it simply dive immediately? How close should a land or sea predator come before breaking cover to give chase? What evasive maneuvers is the prey capable of? Should the prey be herded in a particular direction (such as toward a cliff or other members of the hunting party)? Assuming both predator and prey are in motion, how much "leading" does the target require to assure interception? Air-to-water attacks (such as birds hunting fish) or water-to-air attacks (such as the archer fish, of which *toxotidae chatareus* is an example) must also

An undefined problem has an infinite number of solutions.
Robert A. Humphrey

account for *refraction* (bending of light at the water's surface) that makes objects appear to be in different locations than they really are.

What are the odds of the prey being able to outrun or outfight its attacker? Zebras, gazelles, wildebeests, humans, and more band together into groups or herds that can defend against predation by increasing the odds of any individual surviving the attack. They may also be able increase the odds of a successful defense. Speaking of groups, which individual(s) should the predator target? The closest prey animal may not be the best target if it's in prime health. Seeking those animals that stand out from the crowd (such as moving more slowly, limping, etc.) may increase the odds of success.

A significant amount of computational effort must go into planning every potential attack in order to maximize its chances of success. Attacks are costly, both directly (time and calories invested) and indirectly. The indirect cost or *opportunity cost* of a failed attack can be very high, because that cost includes both the resources invested in the failed attack and the resources that could have been gained with a successful attack.

Attack

Launching an attack places predator and prey in a life-and-death struggle. No prey animal is going to acquiesce to its fate without a struggle, and these struggles are often successful. A lion has a 13% chance of catching a reedbuck (an elk-looking animal), about a 19% chance of catching a gazelle, zebra, or wildebeest, and almost even odds of catching a warthog. Overall, lions have about a 25% chance of catching prey. Hyenas are successful some 40% of the time. Cheetahs do better; their success rate is about 50%, which seems low until one realizes that speed degrades maneuverability. For example, a fighter jet has a much larger turning radius than a small propeller airplane. (I am certainly not suggesting pitting the latter against the former.)

Many animals, humans included, respond to sudden movement. This is why so many nature books admonish readers to move very slowly when confronted by potential predators, such as bears and mountain lions. This is also why humans in

Attacking is the only secret. Dare and the world always yields; or if it beats you sometimes, dare it again and it will succumb.
William Makepeace Thackeray

tense situations with other humans are usually well-advised to move very slowly and calmly. Sudden motion triggers predatory instincts.

Kill

Catching prey isn't the end of the story, because the intended victim may be able to wriggle free, stand and fight, or even have others come to its aid. A giraffe can kill a lion with a well-placed kick.

Most predatory animals are hard-wired to perform a *killing bite* to dispatch their prey. Every species has its own way of performing the killing bite. Examples include shaking smaller prey to death, or clamping onto a larger animal's neck to kill it by suffocation. This bite is usually a quiet affair, and the freshly-killed prey often looks eerily intact afterward. Brain scans of predator animals show that they feel no rage when killing prey. On the contrary, killing for food is a pleasurable experience. And why not? Human anglers delight in catching fish, and human hunters enjoy killing many species of animals.

One does not learn how to die by killing others.
Vicomte de Chateaubriand

The entire predatory process starts with physical hunger that creates emotional urges to find food. These urges give way to the pleasurable anticipation of eating, the thrill of the chase, the joy of the kill, and that certain special contentment that only a full stomach can provide. Herbivorous animals, such as the monkeys and apes we evolved from, experience joy at finding edible fruits and vegetables, and full-belly contentment. And you? Pay attention to your feelings the next time you enjoy a good meal (especially if it comes after exceptionally acute hunger). That sense of utter satisfaction you feel is the same emotion felt by all animals lucky enough to secure a meal.

Coping

How does your average prey animal cope with the fact that its very existence makes it a target for murder? We've already seen some of the ways prey animals can defend against predators; here are a few more of the tricks we would-be meals have up our sleeves.

If a problem has no solution, it may not be a problem, but a fact—not to be solved, but to be coped with over time.
Shimon Peres

Asymmetry

Take a careful look in the mirror sometime. You may notice that one of your eyes is higher than the other. You may also notice that one of your ears is higher and further forward than the other. This asymmetrical arrangement is no accident. Try this simple experiment: Sit motionless in the middle of a room and close your eyes. Have a friend move silently around the room, making occasional noises (such as snapping their fingers). You should be able to tell the direction most noises are coming from, and should have no trouble at all with noises directly in front of and behind you.

Keep experimenting until you hear a noise that you can't pinpoint as coming from in front or behind you. Remain still but open your eyes when this happens to see where your friend is. Chances are that s/he is standing several degrees off-center, either in front or behind you. The imaginary line between your friend and your head is the axis along which sound reaches both of your ears simultaneously. This axis is off-center to allow you to detect and respond to noises directly behind you. Likewise, the slight asymmetry of your eyes enhances your 3D vision.

Asymmetry is a valuable survival tool, but humans nevertheless regard symmetry as attractive. Chapters 19 and 21 discuss this seeming paradox in more detail.

The ability to delude yourself may be an important survival tool.
Jane Wagner

Built-In Fears

Chapter 6 discussed the ideas of built-in fears. It is much easier to be afraid of a snake than a car, even though the chances of dying in a traffic accident are much higher than the chances of dying from a snakebite. One way of interpreting this is that humans are inherently prepared to fear snakes (what Temple Grandin refers to as a *prepared stimulus*).

Reflexes

What man does not understand, he fears; and what he fears, he tends to destroy.
Unknown

Chapter 6 also discussed the presence of both fast and slow fear in humans. Fast fear gets us moving before the slow fear makes us aware of what we're running from. Short-circuiting our conscious reasoning and emotions in this way helps keep us from hesitating and asking questions, which takes valuable time that could be better spent running away.

Causality

Humans and animals have a built-in bias that predisposes us to believe that two events occurring close together are linked—specifically, that the first event caused the second event to happen. This *causality bias* can be a valuable survival tool. It can also lead to superstitions. One day, someone suffered an unfortunate incident soon after a black cat crossed in front of her or him. The superstition about black cats and bad luck was born.

Tell us your phobias, and we will tell you what you are afraid of.
Robert Benchley

Confirmation Bias

Chapter 5 showed you the key role that core beliefs play in building each person's inner reality. Again, core beliefs filter our perceptions before we become aware of them. This process helps contribute to *confirmation bias*, which predisposes us to seek and/or interpret information in a way that conforms to our core beliefs. Confirmation bias also leads us to reject information that could contradict our core beliefs. From an evolutionary standpoint, core beliefs embody our most basic survival instructions. Circumstances that force us to change those core beliefs can therefore be extremely traumatic.

Predicting the Future

Weather forecasts base their predictions on past weather patterns. In nature, past behaviors and past patterns are fairly reliable predictors of the future, and we are biased to recognize this phenomenon. The problem is that humans can apply this bias to situations where the past does not dictate the future. Flipping a penny and getting heads neither increases nor decreases the chance of getting heads or tails on the next flip.

We are made to persist. That's how we find out who we are.
Tobias Wolff

Casinos take advantage of this bias to encourage larger bets. Many roulette wheels feature a billboard that displays recent results, encouraging gamblers to place bets on numbers that haven't come up yet. Who benefits? Not the gamblers!

Fear plays into this bias as well. A situation that resembles a prior dangerous situation can trigger fear, which motivates us to recognize and avoid the current situation.

Universal Distress Noises

There are a few noises that universally communicate or cause fear in most (if not all) animals, including:

- High-pitched sounds. Animals emit high-pitched sounds when distressed, and hearing a high-pitched noise will cause fear in animals that hear it. Intermittent high-pitched noises (such as sirens or some alarms) can be particularly distressing.

- Hissing noises such as snakes and air leaks can cause fear.

- Unexpected mechanical noises, such as clanging metal. The sound of a shotgun being pumped will cause just about any human to pause and reflect, whether or not they know what the sound is.

> *If you are distressed by anything external, the pain is not due to the thing itself, but to your estimate of it; and this you have the power to revoke at any moment.*
> Marcus Aurelius Antoninus

Good Night

The average human spends about a third of her or his life asleep. This means that a person will literally be asleep for 25 years out of a 75-year lifetime. Different animals spend varying percentages of their time asleep. For example, house cats can sleep up to 18 hours per day. In general, predators eat few large meals and engage in relatively short bursts of intense activity when hunting. They also tend to sleep more deeply than prey animals, presumably because they have a lot less to worry about. Herbivores tend to eat frequent small snacks, sleep a lot less, and engage in longer periods of less-intensive activity for foraging.

Sleep allows animals and humans to digest food, remove accumulated wastes from muscles, build up energy for the following day, and perform both routine maintenance and healing/repairs. Sleep also serves to keep animals out of danger when conditions aren't ideal. Human night vision is poor compared to many nocturnal animals, and our predators (cats, snakes,

> *When I woke up this morning my girlfriend asked me, 'Did you sleep good?' I said 'No, I made a few mistakes.'*
> Steven Wright

and eagles) are all diurnal. Sleeping at night is therefore ideal for people.

It follows that most diurnal animals will sleep longer during winter, when days are shorter and food is not as abundant. This pattern reaches its peak in bears and other animals that hibernate all winter. Humans evolved in the tropics, where the amount of daylight remains relatively constant throughout the entire year. Those of us who live in temperate and polar regions tend to be more prone to Seasonal Affective Disorder (SAD), which causes depression during the short winter days.

What Me Worry?

Predator avoidance is our top priority, because getting killed and eaten makes it impossible to do anything else. Prey animals, such as humans, face constant threats, and have few truly safe havens. Could this explain why humans seem predisposed to negativity, worry, and depression? Negativity is all around us, and our vocabularies contain far more negative words than positive ones.

Worry a little bit every day and in a lifetime you will lose a couple of years.
Mary Hemingway

I will say this yet again: I believe this occurs because humans are prey animals who must always be vigilant and worried about the future. This worry helps keep us alive, but can also be traumatic, more so than an actual incident. Soldiers who have been on combat patrols all report that the waiting and uncertainty are worse than any actual fighting. I can only surmise that this holds true for all prey animals.

Chapter 8

Programming the Human Animal

Does human nature stem from our genes, or from our upbringing? Is there such a thing as "normal" behavior? Did dogs domesticate humans? This chapter answers those questions as it presents a brief overview of a few of the many traits humans have in common, how we learn, some of the key limitations on that learning, the importance of play, and a few of the reasons why some conflict is inevitable in daily life. I'll explain some of the evolutionary reasoning behind these topics, and how each evolved to keep us from being killed and eaten by hungry predators.

Nature or Nurture?

Whatever you are by nature, keep to it; never desert your line of talent. Be what nature intended you for and you will succeed.
Sydney Smith

This debate has raged for centuries, if not longer: Is there such a thing as an intrinsic human nature, or are humans products of their environments? Jean-Jacques Rousseau argued for the former during the 1700s, theorizing that humans are born with innate abilities and hard-wired predispositions that are both innocent and good, making babies "noble savages." John Locke argued the opposing view, that humans are born as blank slates with neither knowledge nor skills. Which one of these viewpoints is correct?

It turns out that Rousseau and Locke are both correct. Humans are, like any other animal, born with a full set of genetically coded instincts that predispose us for certain personality traits. Put another way, genes provide the starting conditions that make it easier for certain personality traits to develop. Environment shapes how these traits develop and manifest themselves (including conditions such as clinical depression).

Think of a vehicle manufactured for certain purposes (such as driving off-road) as nature, with the driver who guides that vehicle's actions (on or off-road) as nurture. One cannot separate the two, because the driver is constrained by the vehicle's capabilities and limitations, while the vehicle is obliged to follow its driver's directions to the best of its built-in capabilities.

> *I think people don't place a high enough value on how much they are nurtured by doing whatever it is that totally absorbs them.*
> Jean Shinoda Bolen

This example also applies to humans. Nature and nurture are inextricably linked. There is indeed such a thing as "human nature" that consists of mechanisms perfected over millions of years of evolution (and arguably thrown into chaos by the rapid rise of civilization). There is also a tremendous variety of ways in which this nature expresses itself (such as cultural norms and taboos). In short, all humans have built-in survival instincts. How those instincts manifest themselves on an individual level depends on a combination of genetic predisposition and upbringing.

All humans are human with a common evolved nature; however, differences in genetics and environment can, as we know, yield a tremendous variety of personalities. Let's look at both how human traits vary, and at some of the many traits all humans have in common.

The Bell Curve

Understanding the great variability of human behavior requires an understanding of basic statistics. The image on the following page shows a standard bell curve or *Gaussian distribution* of probabilities. At its most basic, a bell curve graphically displays the likelihood that any given value in a range of possible values will appear. The curve's center is the *mean*, or average of all possible values.

> *A reasonable probability is the only certainty.*
> E.W. Howe

The *standard deviation* (SD) determines how "normal" or "abnormal" a given value is. Values that are within +/- one standard deviation from the mean occur approximately 68% of the time. Values between +/- one and two standard deviations occur approximately 28% of the time. All values exceeding +/- two standard deviations occur approximately 4% of the time.

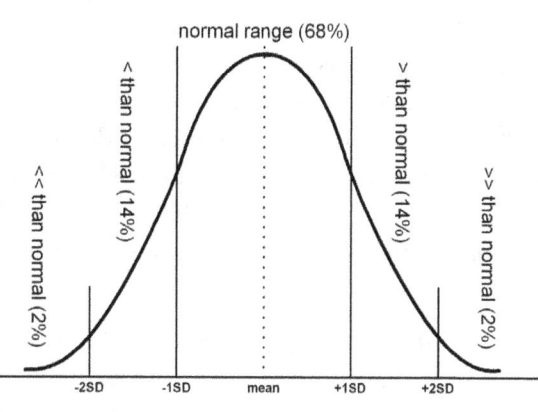

Let's use 100 coin tosses as an example. All else being equal, the mean number of times one can expect heads is 50. In this example, the standard deviation is 5. This means that tossing a coin 100 times will yield between 45 and 55 heads about 68% of the time. You can expect to receive between 40 and 60 heads about 96% of the time, with all results outside these values only occurring about 4% of the time. You can expect to receive between 40 and 45 heads or 55 and 60 heads about 28% of the time (14% each).

> *Do not expect to arrive at certainty in every subject which you pursue. There are a hundred things wherein we mortals... must be content with probability, where our best light and reasoning will reach no farther.*
> Isaac Watts

What does this have to do with human behavior and personality traits? The old adage that there is no such thing as normal is a little closer to the truth than some might like to admit. Back to the coin toss example, the chances of getting exactly 50 heads in 100 tosses of a fair coin is about 8%. The chances of getting exactly 500 heads in 1,000 tosses of that same coin shrink to about 2.5%. There are over 7 billion people on the planet, and a wide range of possible values for any measurable trait. The chances of someone possessing a trait that is exactly the mean are therefore vanishingly small. Measure the same trait in multiple people, and you'll find those measured values scattered around the mean value for the selected trait.

We derive mean values for behaviors and other personality traits by sampling the population at large. Individuals have their own mean values for behaviors and traits that are shaped by genetics and environment, meaning that there is some variability in how the same person may react to different instances of identical stimuli. With that said, the standard deviation for an individual's behavior is much smaller than the standard deviation for a group of people, meaning that the range of a single person's response to stimuli is much narrower than the range for two or more people faced with the same stimuli.

What all this boils down to, is that people as individuals are astonishingly consistent creatures. Old habits do indeed die hard, because consistency is essential for survival. A person is born genetically predisposed to perceive her or his world in a certain way, and life experience codes those predispositions into survival instructions, as discussed in Chapter 5. There is no such thing as "normal" behavior; rather, there is a range of normality for any given behavior—and indeed for any human trait.

This variability is a tremendous evolutionary asset, because it helps guard against the possibility of the entire species going extinct due to a single environmental change. Humans range across the entire planet in every possible environment, each with its own opportunities and dangers. We owe our survival as a species to the great range of possible traits and behaviors open to us.

Canine Connections

Which non-human primate hunts in coordinated groups, weaves complex yet amazingly stable social structures, has plenty of same-sex and non-kin friendships? None. Humans are unique in these regards. Which familiar land mammal species possesses all of these traits? Here's a hint: This mammal is often referred to as man's best friend. DNA evidence indicates that dogs and wolves began separating about 100,000 years ago. Over time, dogs became more and more juvenile in appearance and behavior, because humans tend to favor youth. We'll discuss this more a little later.

Cats are smarter than dogs. You can't get eight cats to pull a sled through snow.
Jeff Valdez

Domesticated animals have one thing in common with their wild cousins: Smaller brains. A dog's brain is about 10% smaller than a wolf's brain. This makes sense, because the brain accounts for over 20% of resting metabolism in humans, and similarly large percentages in other animals. Domesticated animals tend to have reliable food supplies, protection from predators, and outside restrictions placed on such things as where and when they can and can't go places, and when they reproduce. With so many of life's uncertainties and decisions removed, one simply does not need such a large and expensive brain any more than one needs a supercomputer to perform basic arithmetic.

If humans domesticated wolves only to adopt many wolf-like habits (including some necessary for the rise of civilization) then one could say that humans and wolves domesticated each other. One could also expect human brains to shrink as a result of this cross-domestication. It may therefore interest you to know that modern brains are about 10% smaller than Neanderthal brains!

Rhythm

> *You can not evade quantity. You may fly to poetry and music and quantity and number will face you in your rhythms and your octaves.*
> Alfred North Whitehead

Drums. I have yet to find any modern or past culture lacking some form of rhythmic percussion. Rhythm is clearly an extremely important part of human existence. Slower rhythms tend to produce a calming effect, while rapid rhythms tend to produce excitement. This innate reaction occurs in all humans from birth—and possibly even before birth.

One of the loudest noises a human fetus hears is its mother's heartbeat. Faster heartbeats correspond to excitement, exertion, and/or stress, and the infant may even make a Pavlovian association, thanks to chemicals in the mother's bloodstream, such as adrenaline. They certainly make the association in their own lives, as their own heartbeats respond to changing needs. This steady rhythm of life produces about 2.75 trillion beats during a 75-year life span at an average rate of around 70 beats per minute.

80% of mothers cradle their children on their left (heart) sides to comfort them. Traumatized people often rock themselves at about 70 beats per minute, and much of the affection/comforting contact between humans includes rocking movements. Rhythm seems to be hard-wired into us. It has the capacity to soothe and excite, both of which can be essential for survival.

Aww!

Most people would probably agree that puppies, kittens, ducklings, fawns, and cubs are cute, or at least cuter than dogs, cats, ducks, deer, and bears. People also tend to regard babies as cute or at least cuter than children, teens, and adults. Someone who pinches a baby's cheek while making goo-goo noises arouses little reaction, if any, aside from at least tacit approval—even when the baby is not a relation. Pinch an adult cheek while making goo-goo noises, and you'd best have a good explanation if you want to avoid ridicule and possibly even assault charges.

Humans have a built-in protective instinct that is triggered by *neonatal* (juvenile) features in both humans and animals. This instinct is certainly strongest for our own children, less so for related children who are not ours, and least for non-related children. This stands to reason, because the whole game of life revolves around passing on our own genes. Still, it is the rare person who could pass a baby crying alone in the street without stopping to try to help—or a lost puppy, for that matter. I think it's safe to assume that a person confronted by both a crying child and a lost puppy would help the child first.

This makes perfect evolutionary sense. Of course we will protect our children, because they represent the winning home run in the game of life: our genes, passed on to the next generation. We will also tend to protect our kin, because they carry a portion of our own genes, making them an evolutionary base hit. And, given the choice, we will normally opt to help human genes in general pass along by helping a non-related child. After all, a bunt beats a strikeout.

I'm a godmother, that's a great thing to be, a godmother. She calls me god for short, that's cute, I taught her that.
Ellen DeGeneres

There is another potential aspect to the "protect the young" instinct: Humans—even adult humans—retain plenty of neonatal features when compared to other animals. Our bodies are more gracile than those of our more robust ancestors. We are relatively hairless compared to the vast majority of mammal species, and our heads are positively humongous. Babies' heads are larger than adult heads as a percentage of overall body size. We may recognize and want to protect the infant in all of us, absent other pressures, such as reproduction and competition for food and territory that we'll explore later.

Big Head Narrow Pelvis

You may have noticed that women tend to have wide hips compared to the rest of their torsos, resulting in a distinctive hourglass shape. The hips widen rapidly during puberty, which also happens to be the time when human females reach sexual maturity and become biologically capable of bearing children. This is no coincidence. It is also not a coincidence that chimpanzee birth is a straightforward affair requiring less than half an hour to complete, while human labor can continue for hours, during which the emerging infant must rotate more than once to position its head and shoulders. These rotations mean that a human infant normally emerges facing its mother's back, virtually guaranteeing that the mother will need some help during delivery, because attempting to guide the baby out solo risks damaging the spinal cord.

Two competing evolutionary trends make these natal acrobatics necessary. First, humans have big heads. Second, the human pelvis has rotated and narrowed to allow bipedal motion (walking upright). This evolutionary tug of war has had some interesting effects, such as:

- A woman's hips widen at puberty to give her large-headed offspring room to emerge from her womb.

- Humans mature more slowly than our evolutionary cousins. Human and chimpanzee gestation periods are about the same (9 months in humans and 8-9 months in chimpanzees), but human fetal development is slower. A newborn's skull is somewhat flexible, both because the bones haven't finished developing, and because of the *fontanelle* opening in the skull that allows even greater flexibility. Only later in life does the skull grow closed and harden. This flexibility is what allows the skull to pass through the vagina, and it's still a tough fit, as any mother can attest. A newborn's entire skeletal system is extremely flexible, and bones tend to flex instead of break (within limits). This allows the rest of the body to emerge from the mother.

- Humans retain traits commonly associated with baby apes as adults. This *neoteny* (retention of juvenile traits) results in decreased body hair, a larger round head, flat face, in-line big toe, and other conditions such as adult lactose intolerance. A big head allows a bigger brain.

- Humans require parental support long after weaning. While a weaned chimpanzee can forage on its own, a human child would be in serious danger of starvation without a caretaker. As of the first draft of this book, my

---

As we look deeply within, we understand our perfect balance. There is no fear of the cycle of birth, life and death. For when you stand in the present moment, you are timeless.
Rodney Yee

Your birth is a mistake you'll spend your whole life trying to correct.
Chuck Palahniuk

son Logan was almost six years old. He could help himself to whatever is in the kitchen, and would gladly forage for blackberries in season, but I highly doubt he could have survived for any length of time without outside help. A six-year old chimpanzee would no doubt find this puzzling.

- Our slower development and relative immaturity at birth may leave us much more open/vulnerable to initial mental programming that occurs through experience and sets the stage for the rest of the child's life, as we discussed in Chapter 5.

How will the struggle between head and pelvis play out over the course of future human evolution? I won't be around to see it, but I can't help wondering.

Learning 101

A person learns when her or his experiences modify behavior. For example, a child who is bitten by a dog may avoid dogs for many years to come. The experience of being bitten has modified the behavior of approaching dogs; learning has occurred. At least that's how it worked for me.

That is what learning is. You suddenly understand something you've understood all your life, but in a new way.
Doris Lessing

An article in the September 2007 *Archives of Pediatrics and Adolescent Medicine* (Froelich, et. al) reports that about 9% of children have Attention Deficit Hyperactivity Disorder (ADHD). Interestingly enough, the symptoms of ADHD (such as impulsive actions and ease of distraction) may have conferred evolutionary benefits. Someone who is easily distracted may spot danger sooner than someone whose attention is focused elsewhere. Someone who acts impulsively may respond (fight or flight) more quickly than someone who deliberates.

Humans did not evolve to sit in neat little rows for hours on end and to learn concepts from lectures and books that have zero apparent applicability to daily life. There exists a strong correlation between how easily someone learns a task and how applicable said task is for that person's life. The average child learns to talk without any formal training, because communication is a critical survival skill. She learns to walk, run, climb, and feed herself with little help, because these are also critical survival skills. Knowing when George Washington crossed the

Potomac may be valuable under certain conditions for certain purposes, but I highly doubt that any contemporary person's life has hung on knowing that tidbit. (I don't know, nor do I want to know.)

Children are natural-born learners, who actively seek out knowledge with a gusto that few adults can ever hope to rival. To paraphrase Joni Johnston, children seem to intuitively seek out experiences and create adventures that help them make the most of their biological potential. According to Charles Piaget, children are scientists who begin exploring word as soon as they are born, and mental development (emotions) arises from this exploration. Newborns have been likened to pre-programmed friendly computers, the perfect blend of Rousseau and Locke's theories. Babies are students exploring everything within their abilities, with their parents being their most important teachers.

A little learning is a dangerous thing but a lot of ignorance is just as bad.
Bob Edwards

Babies are born lacking a vast number of behaviors that children pick up over time. This does not necessarily mean that such behaviors come from experience. Evidence indicates that behaviors switch themselves on when needed, and that there are certain windows of time when children are predisposed to learn certain things. Absence of behavior at birth does not prove that it comes from experience. Behavior asserts itself when needed. For example, children who are not exposed to language when very young (from birth) may never learn to communicate fully. Victor of Aveyron was a child who grew up on his own until his capture in the year 1800 at age 12. He was never able to learn to speak normally. There are other similar cases. Sexual behaviors (or sexual precursors such as dating) normally don't appear until puberty, when the body becomes biologically capable of rearing children. My son Logan didn't learn to feed himself until he had mastered walking and talking, and had been fully weaned. Children are wired to learn what they need to learn, when that knowledge is needed.

As for ADHD, recent research proves that children with ADHD function better when they spend part of the day out in nature. The more natural and wild the location, the more behavior improves. Forests and meadows work better than parks, which in turn work better than paved surfaces or indoor locations. ADHD aside, muscles tire from sustained exercise and need rest. The human brain tires from sustained concentration of the type expected at school and work, and needs to rest by taking breaks. Breaks taken in nature do a far better job of restoring the brain's ability to concentrate than artificial diversions like games, TV, movies, etc.

This should come as no surprise to anyone. Humans evolved in nature and the rise of cities, civilizations, schools, and jobs has been far too recent for our brains to have adapted. Civilization keeps advancing faster and faster, leaving our primordial minds struggling ever harder to cope. The good news is that technology is slowly making the traditional office obsolete. The day may come when most workers adjourn to the park or a nearby natural area with a computer to get their work done. I do it already, and it's wonderful. As for education, Logan and I enjoy long talks about nature, science, life, and more during our long hikes in the mountains. The depth of his questions and insights never stops amazing me. I can only hope that the research I mentioned above eventually becomes as obvious to educators as it is to me.

Methods

Children learn much through training and imitation, but also have a powerful urge to learn and explore on their own as well. This thirst for knowledge continues throughout life, the fact that you're reading this book being testament to that fact. People do, however, vary in the types of training and other input they gravitate towards. For example, visually oriented people learn by seeing, while those with more kinesthetic tendencies learn by doing. These are just two of the nine types of intelligence proposed by Howard Gardner (see Chapter 4).

Vigorous let us be in attaining our ends, and mild in our method of attainment.
Lord Newborough

Motion

Death Valley's Racetrack Playa is home to a unique phenomenon: This dry lake bed is covered with rocks that show signs of moving across the mud, sometimes for hundreds of yards, and in shifting directions. To date, no one has ever actually seen these rocks move. Logan's first movements were similar. To the best of my knowledge, he could neither walk, crawl, nor roll over when we first brought him home from Korea. Nevertheless, his mother and I would set him down and

Never confuse movement with action.
Ernest Hemingway

return a few moments later to discover him in a different position from where we'd left him. He took off at top speed as soon as he learned to crawl, mastered running in seemingly record time, and now enjoys spending the entire day outside riding his bicycle at full tilt (always wearing a helmet) or hanging out at the playground. My younger brother was the same way, as are all the children I've ever seen.

Children have an innate desire to move, and move vigorously. This exercise builds muscles, bones, and develops gross and fine motor skills and coordination. I see it as no accident that playgrounds provide opportunities to mimic movements needed for survival such as climbing, balancing, hiding, swinging, and sliding. Those with the best physiques and abilities were best able to find food and evade predation. Small wonder that children naturally seek to develop those skills. Evolution most certainly didn't have television, video games, and empty sugar/fat calories in mind for us.

I fully believe that many of society's ills could be solved by giving kids of all ages ample opportunities to exercise in park-like settings. Imagine a large park in each city, combining forested and grassy areas with playground equipment scattered throughout it, with children of all ages encouraged to spend at least a couple hours per day there. I suspect that obesity and ADHD would be rarities. After all, one does not see many fat or "mentally ill" people in primitive societies, at least not as far as my research can tell.

Language

I don't mind what language an opera is sung in so long as it is a language I don't understand.
Sir Edward Appleton

Humans seem to be innately wired for language, possessing what Noam Chomsky calls a *deep grammar* or predisposition toward recognizing the need for nouns, verbs, adjectives, adverbs, and the like to communicate different and often complex ideas. Language is so essential for both group cohesion and predator avoidance/protections that it deserves its own chapter. Please see Chapter 9 for more about language.

Emotions

Cherish your own emotions and never undervalue them.
Robert Henri

I said this before in Chapter 5, and I'll say it again: Humans are emotional creatures whose logic exists to justify said emotions. Babies recognize emotions when they are only a few months old, with empathy being one of the earliest emotions recognized. This stands to reason, because a baby's life depends on caregivers. I know when someone is doing something because they genuinely care about my well-being, and I can only imagine the same holds true for babies.

Babies also use emotions to collect information and make decisions by watching how others act and react to its own actions. This provides critical clues as to what's important versus unimportant, what behaviors to enhance and what others to eschew, what to seek out in life and what to avoid, its own value (self-esteem), and what constructive or destructive behaviors and which emotions are most likely to elicit positive responses from others (or at least the fewest possible negatives). Thus begins the process of realization discussed in Chapter 5 that will color the child's entire life unless she or he takes concrete steps to alter that core emotional programming, such as those described at length in *The Enlightened Savage*. In some cases, failing to achieve certain emotional milestones can impact learning how to speak, read, etc. In short, healthy emotional development must occur to prevent potentially serious learning disabilities. One estimate holds that 15% of the United States population (about 45 million people) has some form of learning disability.

Cheater Detection

Here is an interesting human quirk: Remember the grade-school exercises involving trains leaving stations at certain times and speeds? How about others that ask you to determine, say, the meaning of a fictional word given some contextual information? Many people have a hard time with these types of logical problems, because survival doesn't require that kind of knowledge. I have already argued that the kind of concentration and focus these problems require could even be counterproductive, since they may render the thinker oblivious to danger. Rest assured that these exercises would be second nature to us, had our survival depended on knowing the exact time two lions would reach our location after springing from behind two bushes a given distance apart at given speeds.

A thing worth having is a thing worth cheating for.
W.C. Fields

There is one exception to this general rule: We can easily solve the exact same types of logic puzzles when they are framed in terms of detecting or catching cheaters. This makes strong evolutionary sense. Cheating (such as issuing fake predator warnings or hoarding food) conveys advantages to the cheater at the expense of the group. It therefore stands to reason that an individual has incentive to both cheat and be on the alert for other cheaters.

Cheating and cheater detection can also occur on more subtle levels. For example, if Person A has too much food to eat before it spoils and gives some to Person B, then one could reasonably expect Person B to return the favor when the tables are turned. If Person B fails to do so, then Person A may be far less likely to help B in the future. Similar examples abound in relationships, friendships, and alliances. Fair play is a strong human instinct, which may explain why so many of our laws revolve around the myriad forms of dishonesty.

Security vs. Independence

> *Life is either a daring adventure or nothing. Security does not exist in nature, nor do the children of men as a whole experience it. Avoiding danger is no safer in the long run than exposure.*
> Helen Keller

Babies form attachments to their primary caregivers, and fear both separation from those caregivers and strangers at about six months of age. This is also the age when they are learning to crawl and move on their own, meaning that they are capable of getting lost—a mortally dangerous situation for a helpless baby prey animal. More years must pass before the child's ability to feed and otherwise fend for itself catches up with its ability to move.

This creates a dilemma for both parents and children. Sheltering or over-protecting a child can cause serious developmental and emotional problems. On the other hand, withholding security from children can also lead to serious problems. Parents must walk the fine line of letting go while remaining available at need. The need for independence eventually overcomes the need for parent-provided security.

This is not to say that humans stop needing security. In fact, all people need some sense of security, a sanctuary from the relentless exposure to predation and life's many other demands. This need can only be filled by other people who can provide love, friendship, attention, and protection. Humans are deeply social animals by nature, and the prospect of being alone terrifies most of us (as in forced loneliness, not the quiet solitude most of us enjoy from time to time). The fear or isolation and helplessness in a dangerous world is a fundamental fear that traces its roots back to the time when being hunted was a daily threat.

Children who receive the security they need from their parents while being allowed to explore their world grow up able to give and receive security in their later relationships. Those who don't grow up fearful and deeply mistrustful of others. This is in perfect accord with the process of realization described in Chapter 5.

Limitations

Innate curiosity drives a lifelong thirst for knowledge. Still, there are limits to what humans can learn, as well as how and when they can learn different things. Here are just a few of the key constraints on human learning.

Argue for your limitations and sure enough, they're yours.
Richard Bach

Timing

I've already mentioned that humans have windows of time during which they are able to acquire certain skills. If a person fails to learn language, locomotion, etc. during these windows, then their ability to master the skill in question may be seriously compromised for life.

Honesty

Cheater detection and avoidance really is a big deal for humans. Young children lack many of the inhibitions of adults, but they are excellent at detecting sarcasm, particularly when it's directed at them. Children who experience a lot of sarcasm from their parents and others close to them may have trouble trusting anyone in the future, and may suffer from low self-esteem and depression.

All men profess honesty as long as they can. To believe all men honest would be folly. To believe none so is something worse.
John Quincy Adams

Consistency

Humans may be extremely consistent creatures, but that overall consistency still leaves plenty of room for seemingly inconsistent reactions to situations. Someone who has had a rough day may respond angrily to a situation that wouldn't ordinarily elicit such a response. Children pick up on conflicting signals from their parents. It's impossible to expect that any parent will be 100% consistent with their children; however, parents must pay careful attention to the signals they are sending to keep inconsistency to an absolute minimum. Consistency fosters security and sets clear expectations by which children can evaluate their own behavior and learning progress.

Beyond messages, a child's needs must be met as consistently as possible. Babies can exhibit pleasure and disgust within 12 hours of birth, and can show frustration when their needs aren't met within only a few days.

Pioneering psychologist Karen Horney believed that the treatment we receive from our parents starts the personality ball rolling. I couldn't agree more.

Education

In the first place, God made idiots. That was for practice. Then he made school boards.
Marc Twain

I cannot stress this enough: Humans did not evolve to go to school for 12-plus years or get a job for 40-plus years. Any parent can attest that babies have a hard time keeping still and quiet, a trait that continues in various forms all the way to adulthood. The traditional classroom setting of desks set in neat rows, with success defined as the ability to regurgitate information that often has little bearing on daily life is about as far removed from nature as one can get.

School forces us to use our brains in ways they weren't designed to be used. There is little to no innate motivation to learn reading and mathematics, because these were not needed for survival in our primordial past. Sitting at a desk is one of the least beneficial activities imaginable. Someone who sits still concentrating on something under his nose is a prime target for a predator who can make the kill before the victim knows what hit him. No wonder so many of us have a hard time in school! Success in school demands that we act in a manner that is totally inconsistent with avoiding being killed and eaten.

One of the biggest problems is that standardized education seeks to breed conformity and loyalty, instead of nurturing innate curiosity. Aristotle once said that all who have meditated on the art of governing mankind have been convinced that the fate of empires depends on the education of youth. Albert Einstein, who discovered the theories of relativity that forever altered our understanding of the universe, said that it is nothing short of a miracle that the modern methods of instruction have not entirely strangled the holy curiosity of inquiry. I believe him.

Play

I've mentioned play before, but it bears mentioning again, particularly as it applies to humans and predator avoidance. Play is an extremely essential part of growing up, and remains important throughout life. All work and no play don't just make Jack a dull boy; it makes him absolutely psychotic. Play develops bones, muscles, coordination, and interpersonal skills that may help avoid or fight off predators. It develops and maintains the friendships and alliances that can help secure status later on in life. It allows youngsters to practice important survival skills in a reasonably safe environment. It provides a respite from the daily grind. And it's fun.

By *play*, I primarily mean physical exercise. Video games and other passive forms of entertainment may develop some puzzle-solving skills and fine motor coordination, at the expense of overall exercise when overindulged. The growing epidemic of child obesity bears witness to this. For best effect, play must include plenty of full-body exertion in natural settings.

> *You've achieved success in your field when you don't know whether what you're doing is work or play.*
> Warren Beatty

Conflict

Traditional psychology thinks of functional families as lacking conflict. A traditional psychologist might label familial strife as dysfunctional, but evolution actually predicts—if not demands—a certain level of conflict within families. Every individual in a family has her or his own reproductive interests, which have to interact with the reproductive interests of everyone else. A child that helps raise a younger sibling may be passing on part of her or his genes at the expense of being able to mate and pass on the whole package.

Conflict may even begin in the womb. A fetus has its own interests to protect, and is therefore more interested in its own survival than the mother's, even if its own survival depends on its mother. Meanwhile, the mother is evaluating the fetus's reproductive potential, and may spontaneously abort the preg-

> *Difficulties are meant to rouse, not discourage. The human spirit is to grow strong by conflict.*
> William Ellery Channing

nancy if the fetus is found lacking. Far from uncommon, *miscarriages* (spontaneous abortions) occur about half the time.

Once born, the baby may still face infanticide at the hands of a rival male or even its own mother. Mothers who kill their babies tend to be young, poor, and unmarried. In other words, they lack the experience, resources, and help normally needed to raise a healthy child that is capable of passing along the mother's genes. The mother's own reproductive potential has been proven by virtue of a successful pregnancy; the child represents a huge gamble that could result in both mother and child losing the ability to reproduce. From an evolutionary standpoint, individuals who fail to reproduce have missed the entire point of living. As horrible as infanticide is, it is nonetheless justifiable when seen from the evolutionary point of view. I am certainly not advocating or defending infanticide; I am merely saying that I understand it.

The Programmed Animal

If debugging is the art of removing bugs, then programming must be the art of inserting them.
Unknown

Chapter 5 discussed how humans form the core beliefs that shape every individual's reality. In this chapter, I pointed out some of the many facets of human nature, beginning with how nature and nurture intersect, which allows great flexibility in the specific things one learns, which in turn enhances survival potential in a wide range of environments against an equally wide array of predators and other dangers.

The great variability of human behavioral and physical traits confers a broad spectrum of possibilities and a robust mix of strengths and weaknesses that can help the species survive calamities such as sudden environmental changes. At least one scientist, William H. Calvin, believes that such environmental changes spurred the most recent portions of human evolution that gave rise to modern *homo sapiens*.

Neoteny, the preservation of juvenile traits by adults, may be an offshoot of the same mutation that caused jaws to shrink and allowed brains to expand. Our loss of brute strength lent new importance to developing the mental capacity to outwit predators that could no longer be outrun or outfought on a one-to-one basis.

These examples, plus the other examples I've included in this chapter (as well as many examples I did not include), all contribute to programming the humans animal to recognize and avoid predators, while maximizing the odds of passing on the all-important DNA to the next generation.

Reflecting on the contents of this chapter and those before it, I am reminded of Shakespeare's *Hamlet*. To me, it is self-evident that humanity is most certainly a wondrous piece of work. How noble in reason indeed, except that our reason is no more or less noble than that of any animal that is just as perfectly suited in its own way to its environment. Our faculties are by no means infinite; on the contrary, they are extremely limited. We may have trichromatic vision, but some animals can see infrared and ultraviolet light, and even detect the Earth's magnetic field. Some animals use sound to form mental maps with a precision that renders human sonar a primitive toy by comparison. The list of animals that can out-run, out-climb, out-swim, and outfight a human goes on ad nauseam. In form and movement, we are clumsy oafs next to the acrobatics some animals employ that put *Cirque de Soleil* to shame.

Godlike? I respectfully submit that man have created likenesses of God in our image, not the other way around (which in no way means that God does not exist). The beauty of the world? Beauty is in the eye of the beholder, and I find much that is beautiful and worthy of awe and reverence in all life, against which humans are one form among many. The paragon of animals? That is a matter of definition. The more I look, the more I come to think that every species of life is a paragon in its own right, because all life has evolved from the first cell to what it is today. Humans are certainly not the paragon of the food chain by nature and only sometimes by technology.

Programming today is a race between software engineers striving to build bigger and better idiot-proof programs, and the Universe trying to produce bigger and better idiots. So far, the Universe is winning.
Rick Cook

What is this quintessence of dust that is humanity? That is the question I am seeking to answer in this and other *Savage* books. The short version is that humans are prey animals whose brains are still roaming the African savannas. What you have just read is but a few of the many traits humans have evolved to remain that one crucial step ahead of our would-be killers.

Chapter 9

Communication

Your paradigm is so intrinsic to your mental process that you are hardly aware of its existence, until you try to communicate with someone with a different paradigm.
Donella Meadows

What has the approximate size, shape, and appearance of a dried pear, and is optimized to gather and resonate (amplify) noises in the frequency range of roughly 250 to 3,000 hertz with a sensitivity less than one billionth (1/1,000,000,000) that of ambient air pressure? If you guessed the human ear, you're correct. Our ears can detect sounds ranging between 20 and 20,000 hertz, with a special ability to gather and amplify sounds in the previously mentioned narrower frequency range. It should come as no surprise to learn that said narrower frequency range corresponds to the frequencies that the human vocal apparatus is capable of producing. We are naturally predisposed to hear ourselves and our fellow humans making noises.

Our hearing ability extends far beyond pitch and volume. We can sense changes in pitch that are as small as 1Hz. We are also adept at hearing changes in both volume and *timbre*, which the American Heritage Dictionary defines as "the combination of qualities of a sound that distinguishes it from other sounds of the same pitch and volume." Timbre is what allows you to both recognize the person making the noise and the intention behind that noise.

This highly refined hearing ability would be for naught without an equally impressive ability to produce sounds.

Why Sound?

Sound has many advantages over other forms of expression. Visual systems are useful, and can even be quite complex. Examples include *semaphore* code (communicating by waving flags) and this book. The problem is that one has to see the signal, meaning that both sender and receiver must be within view of each other and looking in the right directions. Imagine if the only way you could warn a friend of looming danger was to leap in front of him gesticulating wildly. My guess is that overpopulation wouldn't be quite so pressing a problem, because the constant shuffling to see one another and pass information along would give away one's position to predators. Not good.

Smell-based communications are widely used by many species as sexual attractors or to ward off predators. This eliminates the line-of-sight problem, but the intended receiver must still be downwind of the sender for this to work. Also, smell can travel a long way. One whiff, and a predator knows exactly where to begin looking. Moreover, it's possible that a group attempting to converse by smell could cause an "everybody talking at once" effect.

Sound requires no line of sight and is only partially dependent on wind direction. One can lie perfectly still and communicate in a whisper or shout a warning without necessarily placing one's self in harm's way. Given a suitably complex mechanism, one can produce a nearly infinite variety of sounds with the potential to effectively transmit a tremendous amount and variety of information.

I was irrevocably betrothed to laughter, the sound of which has always seemed to me to be the most civilized music in the world.
Peter Ustinov

Why Language?

Humans are social animals that live in groups. Remove the ability to exchange even basic information among group members, and you remove most of the benefits of group living. Examples of essential information include:

- Warning of dangers such as predators.
- Announcing opportunities such as a food source.
- Where and when the group will move.

Neither can embellishments of language be found without arrangement and expression of thoughts, nor can thoughts be made to shine without the light of language.
Cicero

- Aggression and status.
- Sexual availability.
- Locating parents or young.

Throw in the need to adapt to different environments, maintain group cohesion and direction, coordinate hunting and foraging, manage agricultural projects, lead troops, describe and use ever-expanding technology, and you find yourself needing an extremely complex system of communication. You find yourself needing language.

I've already stated my belief that humans are not alone in our use of language in Chapters 3 and 4. Humans do seem to have language capabilities that are far richer and more complex than any other animal species, if for no other reason than our anatomy is uniquely adapted to produce linguistic sounds that can be expressed in writing and reinforced by visual signals, such as posture and gestures, what we call *body language*.

The Anatomy of Language

To be positive: To be mistaken at the top of one's voice.
Ambrose Bierce

Let's look at how humans physically produce language sounds and some interesting qualities of the resulting sounds.

Physical Anatomy

Vocalization starts with air exhaled from the lungs that passes through the larynx, which contains the vocal cords that produce "raw" noise. This noise is refined by the mouth, as structures such as the soft palate, tongue, and lips change shape. Human vocal anatomy has been compared to an organ pipe that can change shape in real-time to produce the incredibly rich diversity of meaningful linguistic units called *phonemes*. The English language has about 50 phonemes.

In *The Mother Tongue*, author Bill Bryson points out that the human throat is unique, because we can't breathe and swallow at the same time. The X formed by our mouth, sinus cavities, esophagus, and windpipe that allows us to breathe through our nose and/or mouth and blow milk out our noses, means that there is always some risk of food or drink going down the wrong pipe. According to Bryson, this occurs because the larynx descends into position when a child is between three and five months old—the period when SIDS (Sudden Infant Death Syndrome) most commonly occurs.

If there really is a connection between the larynx and SIDS, then humans are paying a heavy price for our linguistic capabilities. This cost must have been less than the benefits in the great evolutionary trade-off, or it would never have occurred.

Tonal Anatomy

Musical notes appeal to our sense of hearing because of how we product sounds, according to neurobiologist Dale Purves. The ratios between frequencies used to make vowel sounds correspond with the ratios between the frequencies used to make musical notes. Many thousands of languages have existed throughout human history, and many billions of speakers have uttered them, yet all stem from the same frequency ratios. The lowest two vocal tract resonances, or *formants*, comprise spoken vowel sounds. The first formant falls between 200 and 1,000 hertz, the second between 800 and 3,000 hertz. This is exactly the frequency range that the human ear is best optimized to hear.

Purves and his fellow researchers at Duke University also discovered that the frequency ratios of the first two formants formed musical relationships about 70% of the time. This may help explain why humans love music so much: Human speech is inherently musical. Research is now looking at why we associate music played in a major scale with happiness, and music played in a minor scale with sadness. The suspicion is that this may have something to do with human voice characteristics.

Music is the only language in which you cannot say a mean or sarcastic thing.
John Erskine

Beyond Verbal Communication

Humans transmit and receive most of their information nonverbally, despite the resounding advantages of using audible language. Research in communications reveals that words themselves convey about 5% of the total message. You, dear reader, are therefore receiving only 5% of the possible totality of information I am sending you in these pages, because you can only read my words, and must interpret them however you will. It is entirely probable that your interpretation of my words will differ from both those of other readers, all of which will differ from my original intent to a certain degree. If

The most important thing in communication is to hear what isn't being said.
Peter Drucker

this sounds farfetched, consider the many differing interpretations of the Bible that have led to many dozens of sects within Christianity, or how passages of the United States Constitution are routinely interpreted as having different, even conflicting, meanings.

Consider how the words "I love you" could be interpreted as everything from high praise to a scathing rebuke with relatively minor changes in *intonation*, such as speech, volume, or rapidity. Intonation conveys about 35% of the message, making it 7 times as effective as mere words. Put another way, you'd get about 8 times as much information out of a CD of this book than you are from simply reading it—and no, I'm not trying to up-sell you.

The remaining 60% of information is conveyed non-verbally through posture, gestures, facial expressions, and other subtle cues that many speakers don't even realize they're sending. These cues speak volumes about the speaker's sincerity and intention. Humans recognize the meaning of facial expressions and some other body language cues around the world, regardless of culture or spoken language. Non-verbal communication is universal. This explains why trying to communicate with someone in a foreign language involves a lot of gesticulating. We "get" such communications on an instinctual level.

Language Evolution

It is of interest to note that while some dolphins are reported to have learned English—up to fifty words used in correct context—no human being has been reported to have learned dolphinese.
Carl Sagan

To be effective, a means of communication requires some means for conveying what is being talked about (nouns), what action is taking place (verb), and the ability to describe both the subject of conversation (adjectives) and the action taking place (adverb). These components allow us to exchange information, such as "Big cat bites hard" (adjective, noun, verb, adverb).

Exchanging more complex information requires additional components, such as tense (past, present, or future). For example, "Big cat bit hard" and "Big cat will bite hard" convey two very different meanings than the original sentence. Informing someone about things that have been completed, are continuing, or have some historical interest requires using the *perfect* tenses (past perfect, present perfect, and future perfect). Need to express something uncertain or non-factual? If I were you, I'd use the *subjunctive* tenses. Other parts of speech become necessary for complex communications including pronouns, prepositions, conjunctions, and interjections.

Each of the eight parts of speech has several sub-types such as verb tenses, several types of pronouns (such as indefinite and interrogative), and more. Thus is the grade school hell of diagramming sentences and conjugating verbs on the blackboard born, for without rules this system would be absolutely unworkable, and we would be unable to effectively communicate complex information.

Human language allows (or at least seems to allow, given the current state of our knowledge) much more complex, precise, and efficient communication than non-animal language. I say "seems" because ongoing research is revealing that animals sounds are much more complex than initially believed. Sophisticated electronic analysis is exposing subtle differences in "identical" animals calls that the human ear and less advanced machinery are unable to detect. It seems certain that language is at least partially responsible for the sudden explosion of human innovation and advancement that occurred only few tens of thousands of years ago. Language is most certainly indispensable for modern life. Without it, life as we know it would be utterly impossible.

One ought, every day at least, to hear a little song, read a good poem, see a fine picture, and if it were possible, to speak a few reasonable words.
Goethe

Vervet monkeys have words for different types of predators (they can distinguish predators from harmless species), parent/child recognition (possibly analogous to naming individual people), status, and more. This is arguably a simple language, and goes to show that human inability to detect animal language doesn't mean it's not there. After all, we often can't make sense out of what other people are saying. Captive apes have been taught complex artificial languages, and I must presume that this ability exists in the wild, since there are no inherent fundamental genetic differences between captive and free apes that I know of.

Does animal language include grammar and syntax? That's an open question, and an intriguing one because grammar and syntax provide the framework that allows one to extract the intended meaning from strings of words. In English, reversing "I love you" to "you love I (me)" conveys completely different meanings. Without grammar, both sentences would have the exact same ambiguous meaning, because it would be impossible to tell whether I love you or you love me.

A robust grammatical system allows one to construct complex sentences and convey precise meaning. Grammar is so important, that many human words consist of grammatical constructs with no concrete references. Much of human language describes concepts and contexts in the abstract, as opposed to things in the immediate vicinity at the present time. Grammar also gives rise to hierarchical language by providing rules by which a few dozen building-block sounds (letters) combine into many possible syllables, which in turn give rise to hundreds of thousands of words that can be strung together in nearly infinite ways.

I suspect we'll eventually discover that vervets and other species (especially the great apes such as chimpanzees, bonobos, and gorillas) have amazingly rich languages, complete with some form of grammar. I reiterate my argument that there is no known genetic difference between captive apes who can learn human grammar and wild apes. I must also point out the seeming necessity for fairly extensive grammar and vocabulary using the following examples.

Imagine you're a vervet monkey foraging for food on the ground with the rest of your troop. Someone screams in alarm. This tells you that something is wrong, but what? Did something happen to an individual (such as an injury) or is the entire troop in peril? Now imagine hearing "Predator!" This tells you that you are in mortal danger, but tells you nothing about the threat. Should you run up a tree, or hide under a bush? Now imagine hearing "Leopard!" Now we're getting somewhere, because you know that you'd best be hauling yourself up a tree, which could be the exact wrong thing to do if the predator was an eagle. Vervets are known to have this level of vocabulary, as described above.

This still leaves the potential victims in a precarious situation. Which direction is the leopard coming from? This would be extremely useful to know, lest your escape attempt lead you straight into its jaws. If the leopard is coming from the north, you may want to climb the tree to the south, even if it's farther away than the northern tree.

It is better wither to be silent, or to say things of more value than silence. Sooner throw a pearl at hazard than an idle or useless word; and do not say a little in many words, but a great deal in a few.
Pythagoras

We have too many high sounding words, and too few actions that correspond with them.
Abigail Adams

How far away is this leopard? Mice are known to react very differently when they smell a distant cat (little to no danger) versus smelling a nearby cat (imminent danger). I rather doubt that vervets wait until a leopard is right on top of them before sounding the alarm, just as I doubt that most pedestrians would wait for oncoming traffic to draw near before stepping off a curb. The decision to walk or run to the sidewalk depends on both how far away traffic is and how fast it's approaching. I can only assume that the same logic holds true for vervets confronted by a leopard.

Never part without loving words to think of during your absence. It may be that you will not meet again in life.
Jean Paul Richter

I grant that this isn't always possible. "Leopard!" may be much more effective than, "Leopard, from the north, fifty feet, closing fast!" for obvious reasons. That said, even a few of those extra words can convey potentially life-saving information. I believe that the potential for saving lives is great enough to make it a fairly safe bet that vervet and other animal communications are rife with undiscovered complexity.

The earliest known human writings are every bit as complex as modern languages, which suggests that language must have evolved before then. One might think that primitive societies might be a good place to hunt for primitive languages, but it turns out that these societies have languages that are every bit as rich and complex as our own—if not more so.

Just how do languages evolve? No one can yet answer how language evolved in humans, although the *FoxP2* gene described later in this chapter offers some exciting possibilities. We can, however, describe how languages spring up when people from different cultural and language groups come together, and can also guess what happens before that.

Simple grunts or calls become single words, and the two may already be functionally identical. Next come 2-word sentences (typically noun verb), then multiple-word sentences that lack grammar and seem limited to simple concrete references. Finally, new words evolve that have purely grammatical functions and lack concrete references. Let's see how this works when two peoples speaking different languages encounter each other and need to communicate, such as when English-speaking traders arrived in Melanesia in the 1800s.

Make sure you have finished speaking before your audience has finished listening.
Dorothy Sarnoff

Pidgin is the first step in evolving a new language. A *pidgin* is a simple combination of sounds borrowed from both languages and mashed together in short sentences, with no formal construction and no grammatical rules. If the contact between peoples is short, rare, and/or only occurs for specific purposes, then no further evolution may occur. Sustained contact, especially among large numbers of people, can spur the rapid evolution from pidgin to creole. This evolution happens even faster when one group adopts the emerging creole as a native tongue. A *creole* is a mixture of two or more languages that includes its own vocabulary and grammatical rules.

> *The individual's whole experience is built upon the plan of his language.*
> Henri Delacroix

In *The Third Chimpanzee*, Jared Diamond provides wonderful examples of Neo-Melanesian, a creole of English, Tolai, and other Melanesian languages. Creoles have elaborate grammatical rules, but remain less efficient than fully matured languages. For example, "grass belong seawater" is the Neo-Malaysian term for "seaweed," while "man he got no grass long head belong him" means "bald man." In the above examples, I altered the Neo-Melanesian spelling to standard English to aid understanding, but left the actual words in their original order. In short, creoles string existing words together to fill in the gaps, while mature languages simply invent another word.

I make the claim that all human life boils down to the six elements described in this book, regardless of culture or ethnicity. One culture may appear radically different than another on the surface, but deeper examination reveals these differences to be differences of manifestation only. Humans are humans with the same needs, no matter where they live. The same is true of creoles. All creoles are different just like all languages are different, but all share the same fundamental features when one looks under the hood.

> *Words are the leaves of the tree of language, of which, if some fall away, a new succession takes their place.*
> John French

Linguist Noam Chomsky believes that language is too complex to master in a few years, and that humans must therefore have an inborn "deep grammar." Derek Bickerton takes this idea even further by theorizing that humans are genetically wired with grammatical switches that result in all languages having remarkably similar grammatical features. If this is true, then children should learn the equivalent of pidgin and creole before mastering full language. As we'll soon see, that is exactly what happens.

Learning to Speak

I must confess that I failed 8th grade English at the Lycée Louis Pasteur in Bogota, Colombia, despite having earned stellar marks in English during all of my earlier grades. I'll never forget marching up to the teacher on the first day of class to inform her that I had grown up in the United States, where I'd been speaking perfect English for as long as I could remember. She responded by exempting me from all class work, except tests. I flunked every single one, despite being a native English speaker with strong fluency in both Spanish and French, which included conjugating many verbs in all imaginable tenses.

People come from around the world and can understand each other without even speaking the same languages!
Segei Bubka

French and Spanish share many grammatical rules that are either different, or lack direct counterparts in, English. Furthermore, my French and Spanish education had been rule-based, which explains why I could conjugate any tense on command. By contrast, my English education had been usage-based. I learned how to speak, read, and write without learning the names of every rule. Add a pinch of pubescent arrogance, and the stage was set for academic disaster.

Remember Chapter 8, where I asserted that humans didn't evolve to sit at desks passively absorbing information? My experience learning English was far more natural than my many blackboard hours with Spanish and French. English is what my parents and everyone around me use for daily communications. I therefore had extremely strong innate motivation to learn English. The other languages are nice. I use them regularly. Still, they are far less important, and were much harder to learn. Logan's cousin Noah spent several years in France with his English-speaking parents. He had every incentive to master both languages, and did indeed became fluent in both.

These anecdotes beg the larger question: How do babies learn to talk? Surprisingly, they may do so using the same methods some birds use to learn to sing. According to psychologist Michael Goldstein, interaction plays an important role in language acquisition, because it guides the baby's learning, thus complementing simple imitation. If you've ever gone gaga over a baby saying "dada," "mama," or your name, then you sent that youngster a strong message that helped guide her or his learning. This theory directly contradicts standard thought that suggests speech comes through imitation alone, while suggesting that learning to speak might be much more complex than previously thought.

Language Genetics

There is no excellent beauty that hath not some strangeness in the proportion.
Sir Francis Bacon

Getting someone else to raise your child is *cuckoldry*. Cowbirds have elevated cuckoldry to an art form by laying eggs in other species' nests for the unwitting foster parents to raise. Young cowbirds fatten themselves up for a few weeks before flying off to join whatever nearby flock of cowbirds they can find. Reproductive habits aside, cowbirds are interesting, because females teach males to sing despite being unable to sing themselves. Young cowbirds experiment with different songs, and the females respond when those experiments yield optimal results. It's like a mute person teaching a baby to speak by listening to random babbling, and nodding encouragement when those babbles contain some meaningful snippet. Over time, the meaningless snippets get weeded out.

Experiments with humans mothers and their babies demonstrated that babies who received parental approval when uttering speech-like sounds improved both the amount and quality of their vocalizations, compared to the control group that did not receive this encouragement. Their progress stagnated. These experiments were too small to offer proof of Goldstein's theory, but this does not discredit the intriguing results.

The similarities between humans and cowbirds is no accident. Both species share a common gene that helps shape the ability to acquire language. Sebastian Haesler and Erich Jarvis studied a gene called *FoxP2*. Mutating *FoxP2* causes specific learning disabilities in people, wherein they have normal motor skills but cannot pronounce words, create correct sentences, or comprehend complex language. *FoxP2* levels spike just before a bird starts changing its songs. Activating *FoxP2* wakes up the brain's language (or song) learning circuits.

My guess is that human *FoxP2* levels rise when the baby is about four months old, which triggers such behaviors as intense interest in watching parents and others speak, attempts to "talk," cooing, and the emergence of different cries for different situations (such as hunger, wet diaper, or fright). Infants in the second half of their first year start to understand some common words, and start making coherent noises that sound distinctly word-like. By 18 months, the growing toddler should understand simple questions and babble in sentence-like sequences, as well as nod and shake their heads for yes and no. The ability to string words together in simple noun-verb sentences soon follows, along with the ability to follow simple directions and rapidly expanding comprehension. By age 5, the child can follow compound directions (do A, then B, then C), and can accurately pronounce most words, as well as define some words. *Fox2P* may be the genetic switch envisioned by Bickerton that activates the deep grammar Chomsky believes is part of human nature.

The *Global Language Monitor* estimates that the English language contains just under 1,000,000 words. Differing estimates place the number of different words used on any given day at between 800 and 3,000. Of course, we use some words far more often than others. People talking to children will usually simply their vocabularies even further, depending on the child's age. This "subset" and "subset of a subset" approach may greatly simplify the child's language learning process. Mastering the basics makes learning advanced concepts easier. Logan regularly pronounces new words correctly the first time he hears them, and both requests and remembers their definitions.

Literacy

It has been most interesting to watch Logan's progress in learning to read and write. In the beginning, he read and wrote individual letters and numbers and understood that L-O-G-A-N spells his name, but had yet to grasp that each of these five letters symbolizes a particular sound, and that stringing these sounds together forms his name (EL-OH-GUH-AH-AN).

A book burrows into your life in a very profound way because the experience of reading is not passive.
Erica Jong

Speech comes naturally, but reading and writing don't, because these forms of communication did not become important until very recently in our evolutionary history, when the advent of civilization made record-keeping and formal education important. Logan, like all normal children, acquired spoken language with zero formal training. In fact, he accidentally acquired a little too much language, and had to be taught that certain four-letter words are only appropriate to say at home—and that only under very special circumstances. Learning to read and write has kept him busy for years, just as it did for me.

How many a man has dated a new era in his life from the reading of a book.
Henry David Thoreau

Again, this is perfectly understandable, because literacy was just not a survival requirement on the African savanna. I will even go so far as to speculate that the ability to read and write could be detrimental in the wild, because someone concentrating on translating otherwise meaningless squiggles into coherent language concepts (or vice-versa) is a much easier target than his illiterate—but vigilant—companions.

110 | *The Natural Savage*
Discovering the Human Animal

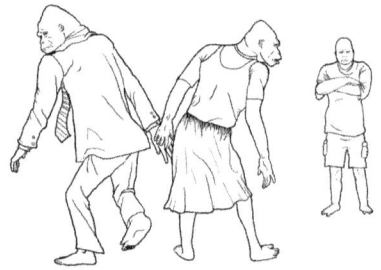

PART THREE

Status

112 | *The Natural Savage*
Discovering the Human Animal

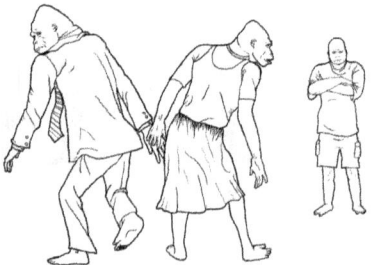

Chapter 10

The High Price of Fitting In

> *We are born charming, fresh and spontaneous and must be civilized before we are fit to participate in society.*
> Judith Martin

Prey animals tend to be more social than predator species, because group living lowers the odds of a predator attacking any one individual, and may also allow the group to repel attacks. A lone monkey has little chance against a predator; a group of monkeys literally has a fighting chance.

This is a very general rule of thumb, and exceptions abound. Some prey animals, such as the orangutan, live alone. Some predator animals, such as wolves and hyenas, hunt in packs. Orangutans are interesting, because they live alone and spend much of their time in the trees. Their chief predators are Sumatra tigers (*panthera tigris sumatrae*) and leopards (*panthera pardus*). Interestingly enough, there are no orangutans in any areas where leopards live. Leopards can catch arboreal prey, and the orangutan's solo existence makes it easy pickings.

I have already argued that humans are nothing more or less than just one species of prey animal on a planet teeming with animal species. Humans are also highly social animals, to the point that group status and the power that comes with that status is one of our six core life functions. I occasionally have to slam on the brakes while driving through downtown San Francisco to avoid a homeless person crossing the street against traffic. These episodes are not the last-moment lunges of a suicidal person; they are slow deliberate movements seemingly designed to cause as much disruption as possible. I used to bristle at these episodes, before it finally dawned on me that these people at the lowest rungs of our society are exerting what tiny amount of control they can over their environment and those around them. To me, this is one more piece of evidence that we will seek whatever power we can by whatever means we have available. Classical psychology tends to ignore this fundamental human need, except in cases of abuse.

The Importance of Society

In evolutionary terms, a lone human is as good as dead. We lack the means to outrun, out-climb, or outfight our predators on an individual basis. The only way enough of us can survive long enough to fulfill life's ultimate goal of reproduction is to band together for the common good.

True to form, most humans crave interaction with other humans. We seek out opportunities to affiliate ourselves with all kinds of groups, from knitting circles to nation-states, and go to extraordinary lengths to make ourselves acceptable to the groups we want to join. The teenager sporting an unsightly haircut and wearing outlandish clothes while listening to music that most adults dismiss as noise is doing nothing more or less than seeking social acceptance among her or his peers—a drive that is rooted in our deepest survival instincts. From our language (official and slang) to the clothes we wear, the food we eat, where we live, who we vote for, the god(s) we pray to, the wars we fight, the customs and rites we observe around major life events, taboos, education, and more, all of us are doing our utmost to fit in. We are all trying to keep up with the Joneses, whether we like it or not.

Society does have its challenges. Some hierarchy is needed to enforce group identity, and set group direction and cohesion. As we'll see in Chapter 11, there are many benefits to being in a position of power and competition, for status exists at all rungs of the ladder. Group living means curtailing individual freedoms in order to enjoy varying degrees of freedom and protection conferred by the group on its members. This is the basis for what Jean-Jacques Rousseau termed the *social contract*. Rousseau wrote (as translated by Maurice Cranston):

Few men are willing to brave the disapproval of their fellows, the censure of their colleagues, the wrath of their society. Moral courage is a rarer commodity than bravery in battle or great intelligence. Yet it is the one essential, vital quality for those who seek to change a world which yields most painfully to change.
Robert F. Kennedy

Chapter 10
The High Price of Fitting In

What man loses by the social contract is his *natural* liberty and the absolute right to anything that tempts him and that he can take. What he gains by the social contract is the *civil* liberty and the legal right of property in what he possesses. If we are to avoid making mistakes in weighting one side against the other, we must clearly distinguish between *natural* liberty, which has no limit but the the physical power of the individual concerned, and *civil* liberty, which is limited by the general will; and we must distinguish also between *possession*, which is based only on force or the "right of the first occupant," and *property*, which must rest on a legal title.

Any person has the intrinsic ability to harm another person at any time. For example, there is nothing physically stopping a dinner guest from lunging across the table and stabbing his host with a steak knife, nor the host from poisoning the food. Under normal circumstances, however, host and guest agree not to harm each other. This is the fundamental idea of the social contract: I promise not to do anything to harm you, provided you do nothing to harm me. This arrangement allows everyone to live longer, more comfortable lives, and may indeed be wired into our brains. Society would disintegrate and take the human race with it but for these powerful inhibitions against harming others.

Here again, there are exceptions. From not returning a favor to murder, all societies have members who don't uphold their end of the social contract. Societies have individual and collective methods for detecting, responding to, and correcting these misdeeds. Remember that humans are extremely adept at solving problems that involve cheater detection, to the point that we can solve a given logical challenge far more quickly when said challenges are presented in cheater-detection terms. The friend who doesn't return the $5 he borrowed as promised will have a much harder time securing a second loan from you. The individual who murders another person receives extremely harsh punishment, up to and including being deprived of her or his own life. Warfare is a special exception to this rule that I'll discuss in Chapter 12.

So far, I am using the term "society" in a very generic sense. It is important to note that group affiliations (societies) can

---
We live in a vastly complex society which has been able to provide us with a multitude of material things, and this is good, but people are beginning to suspect we have paid a high spiritual price for our plenty.
Euell Gibbons

occur at many levels, from informal book clubs to nation-states. Each such group at any level has its own norms, customs, expectations, taboos, etc. It also has its own opportunities for leadership. We'll explore this idea more in Chapter 11.

Civilization

Humans are the only civilized species on the planet by many definitions; however, as we saw in Chapters 3 and 4, there is no natural law that limits civilization to humans, or even hominids or mammals. Given enough time to evolve, it's possible that most any type of animal could achieve what humans like to think of as civilization. Given what we know about the fundamental aspects of civilization, it may be reasonable to conclude that other species have already achieved this.

Critics will rush to point out that humans are the only species with advanced governments, cohesive tribes, art, language, technology, and more. I for one believe that the facts and theories presented in this book counter any such arguments. It's more than a little humbling to step down from our self-made pedestal. This humility in the face of our overwhelming non-uniqueness as a species may result in closer appreciation of our membership in—and intimate connection to—all of nature. This in turn may be the catalyst for the fundamental changes humans must make in both worldview and daily life, if we are to continue our unprecedented success as a species—or at least not take the rest of the world with us when we fall.

> *It was the Law of the Sea, they said. Civilization ends at the waterline. Beyond that, we all enter the food chain, and not always right at the top.*
> Hunter S. Thompson

Selection Pressures

The list of humanity's most successful civilizations spans both history and the globe. From the ancient Babylonian, Egyptian, Greek, Roman, Carthaginian, Minoan, and Persian empires to the medieval empires of Britain, Spain, Portugal, Turkey, Mongolia, and Central and South America, to the modern United States and China, civilizations have flourished across the globe. The two common features of each highly successful civilization are economic and military superiority.

This makes sense. Civilizations with superior economies can out-produce rivals. Those with better armies can dominate by

> *Five senses; an incurably abstract intellect; a haphazardly selective memory; a set of preconceptions and assumptions so numerous that I can never examine more than minority of them—never become conscious of them all. How much of total reality can such an apparatus let through?*
> C.S. Lewis

force. These are roughly the same factors that determine success in animal societies. Groups of any given species that have dependable access to abundant food tend to be larger than groups lacking such resources. Groups better able to compete for territory are better able to take over adjoining territories. Thus goes the endless cycle of competing for the privilege of passing on as much of the all-important genetic material as possible. The individual drive to reproduce as often as possible either directly (by having sex) or indirectly (helping kin survive) all but guarantees that competition must occur between group members and between groups.

Those who have commented on the human condition throughout history are quick to point out that success comes with a high price tag in the form of increasingly destructive warfare, conquest, class and caste systems, and more. The trend is clear: Succeeding generations of human society are marked both by ever-improving technology and by ever more destructive ends. Some estimates place the total population of the Roman empire at about 60 million people. Over 70 million soldiers and civilians died during the Second World War that saw the effective end of the British empire and the rise of the United States of America. The advent of nuclear weapons and other weapons of mass destruction means that the next great struggle for dominance may well spell the end of the human race.

Chapter 12 discusses war in more detail. In the meantime, it is interesting to note that some supposedly primitive societies have managed to solve some of the seemingly intractable problems faced by other more "advanced" societies. Is this thanks to some radically different cultural manifestation of basic human nature? It it because small primitive bands that tend to inhabit sparsely inhabited areas that are rich in natural resources have no need of warfare, violence, or caste? I don't know. I can say that humanity is at a critical juncture where we must rethink our interactions with ourselves and our host planet if we are to continue thriving. Chapters 24-26 outline my thoughts on this topic in more detail.

From Trees to Towers

Humans, like chimpanzees, seem to harbor a deep-seated fear of our fellows. The amount of fear we experience is inversely

---

You can't say that civilization don't advance, however, for in every war they kill you in a new way.
Will Rogers

I have come to realize that all my trouble with living has come from fear and smallness within me.
Angela L. Wozniak

proportional to how close the other person is to us. For example, we tend to fear our blood kin the least, followed by members of our social group (community, neighborhood, coworkers), then members of our tribe (nations), and finally others. Differences can be actual or perceived. A person might fear someone of a different ethnicity who lives nearby than someone of the same race who lives thousands of miles away. Differences in religious beliefs can also foster intense fear. Islam is based on Christianity, which in turn is based on Judaism. The common foundation shared by all three religions is not enough to prevent widespread mutual fear, intolerance, and resulting conflict.

I am fully convinced that anger is a fear response to actual or potential harm. If you've ever felt anger at being insulted, then that anger probably originates in the deep-seated fear that some or all of the insult may be true. If you've ever felt anger at being assaulted or otherwise physically or intellectually challenged, it is because you fear defeat. An insult or other challenge that you either know to be false or where you don't care about the outcome triggers no fear, and hence no anger. My dictionary defines *anger* as a strong feeling of displeasure and belligerence aroused by a wrong. This is perfectly consistent with anger being a response to fear.

What do we fear? Loss of resources and/or status, which negatively impact our ability to reproduce. Who presents the biggest obstacle to our reproductive success? Other members of our species who are not related to us, and/or who may not even be members of our own tribe or society. Nothing can compete with us quite as well as members of our own species, each of whom possesses roughly the same abilities as ourselves, requires the same resources and status in order to reproduce, and is just as eager to get one over on us as we are to beat him.

Train yourself to let go of the things you fear to lose.
George Lucas

Persistent or chronic fear can lead to *hate* (a feeling of extreme aversion or hostility), which is a response to *xenophobia* (an unreasonable fear or hatred of foreigners or strangers or of that which is foreign or strange). Xenophobia is not unique to humans. Chimpanzees certainly have it, as evidenced by warfare between rival groups. I'd go so far as to argue that all animals species that exhibit aggression between social groups also experience xenophobia. Xenophobia-induced murder has a long bloody history.

The only major difference between animal and human xenophobia is that humans are the first species capable of ending all life on Earth. It's ironic how the same competitive pressures that helped humanity flourish may be our undoing.

Xenophobia comes naturally to humans, and our weapons make it deadly. Xenophobic murder has lots of precedents, but humans are first species to be able to end all life on Earth with it. The switch from hunter-gatherer to "civilized" life has only been occurring for the last 10,000 years or so. Our brains are still living in the very distant prehistoric past. The same instincts that helped us leave the primeval forests are alive and well, both in what remains of Earth's jungles, and in the modern "trees" we erect in every major city.

> *Power consists in one's capacity to link his will with the purpose of others, to lead by reason and a gift of cooperation.*
> Woodrow Wilson

Getting Along

"Everything in moderation" is a good rule to live by. Unfettered competition would violate the social contract and make life unbearably difficult if not impossible. We must therefore balance our need for competition with strong cooperation. This cooperation manifests itself in *maintenance behavior*, an elaborate system of rituals designed to keep the peace. Two hikers passing on a wilderness trail are very likely to nod and greet each other. This simple act tells the other "I mean you no harm." The nod is a submissive gesture that appeases the other person by according them status, and the greeting is a mutual peace declaration.

Similar behaviors occur every day, often without any awareness of what's going on. The myriad greetings, asking how someone is doing, wishing someone a nice day, and the elaborate efforts we take to avoid colliding with others are all examples of how humans actively strive to maintain the peace. Bump into someone on the street, and both parties will most probably apologize immediately, regardless of who is at fault. Koreans have taken this peacemaking one step further: Urban population density is so high that collisions are common—so common, in fact, that apologizing every time would put a serious dent in daily schedules. Instead, Koreans assume an implicit mutual apology, and keep right on going. I don't suggest doing this in an American city! I'll expand on this concept more in Chapter 12.

I love to throw parties and invite many friends from all walks of life. The spectacle is always the same: People instantly cluster with those they recognize. New arrivals make the rounds of greeting everyone before settling into a comfortable group for the inevitable small talk. As the tension of meeting strangers on someone else's turf fades, the groups begin to blend and mix with exchanges of small talk giving way to genuine conversations. At the end of the evening, the small talk resumes. This process breaks the ice, establishes rapport, and reaffirms goodwill.

They say that blood is thicker than water. Maybe that's why we battle our own with more energy and gusto than we would ever expend on strangers.
David Assael

Humans are extremely territorial, to the point where we employ highly trained specialists to survey land and tell us exactly where boundaries lie to within tiny fractions of an inch. Some of the many forms one completes when buying a house include disclaimers that fences may not lie on actual property lines. Entering another person's property is cause for stress on both sides. Greeting and welcome rituals allow visitors into homes and businesses. Once in a home, guests ask permission to visit different areas of the house (such as the rest room). It is only once friendship is established that people will feel comfortable moving about other people's homes. The closer the friendship, the greater the comfort. In general, any access granted to visitors extends to the visitors' children, who are treated as extensions of their parents.

Friendship First

I know from personal experience that one can never have too many friends. Friendship is important for both humans and animals. Chimpanzees rely on friends to help them ascend the ranks. Giraffes spend 15% of their time grazing near friends, and only 5% grazing near other giraffes. Human friends can open doors that would otherwise be closed, and can help in many other ways. The end of Chapter 14 talks more about friends.

Never refuse any advance of friendship, for if nine out of ten bring you nothing, one alone may repay you.
Madame de Tencin

When it comes to making friends, the old saying that one never gets a second chance to make a first impression isn't far off the mark. People can often recognize whether a new acquaintance will benefit them or not. We are also extremely adept at recognizing both facial expressions and nonverbal communications across cultures. Interestingly, we have a much easier time picking an angry person out of a crowd of happy people than a happy person out of a crowd of angry people.

Under Pressure

> *No pressure, no diamonds.*
> Mary Case

Primitive hunter-gatherer tribes generally consisted of up to a few dozen people. Everyone knew everyone else, and the entire group ranged over wide areas. Like chimpanzees, clashes between neighboring human tribes were probably frequent enough to make warfare a natural human activity. And why not? Eliminating the competition increased available resources and enhanced the odds of passing on the all-important DNA.

Fast forward to today, where millions of people live cheek by jowl in ever-expanding cities. On any given day, a modern human can expect to see more people than our ancestors did in an entire lifetime. Whether this compression is behind individual and social ills such as mental illness and crime, as Desmond Morris asserts in *The Human Zoo*, is questionable. The human animal has developed extremely effective methods of coping with increased population density; however, the fact remains that the human population is under increasing pressure in both density and overall numbers.

Help!

> *A friend is someone who will help you move. A real friend is someone who will help you move a body.*
> Unknown

If competition for resources and reproductive rights to the point of open warfare represents one extreme of human social existence, then *altruism* (unselfish concern for or devotion to the welfare of others) represents the other extreme. Altruism is widespread among people and animals. Helping one's kin helps pass on at least some of one's own genes. Helping fellow members of our society keeps the group strong and better able to fight off danger or conquer other weaker groups.

There may be no immediate direct benefit to a given act of altruism, and that act may even harm the individual (such as a soldier throwing himself on a grenade to save his buddies at the cost of his own life). Still, this does not mean that altruism is entirely selfless. Throwing one's self on a grenade enables the group to survive. It is highly doubtful that a group of soldiers includes blood relatives; that group does, however, provide mutual protection, and may even stand in for a family group. In other words, at some level, the soldier offering up his own life may believe that he is helping his tribe—and his genes—survive. Remember that groups of people who lack blood ties are a relatively new phenomenon in human evolution.

Self-sacrifice makes sense when seen in this context. If me giving my life means that my genes have a greater chance of surviving, then I have a powerful evolutionary reason to do so. Under these circumstances, my genes benefit far more by my death than by the death of the entire group. I may intellectually know that the people I'm saving aren't my kin but we already know that decisions are anything but intellectual.

Altruism is most likely to occur when the benefit outweighs the cost. This benefit can be direct, such as the status and recognition that come from helping a little old lady cross the street or helping a relative survive by donating a kidney. They can also be indirect. For example, a hunter who killed a large animal and lacks refrigeration may share his meat with other hunters. Here, the cost is negligible ,because the hunter cannot possibly eat the meat before it spoils. The potential benefit is huge, because the hunter may not always be this lucky. Sharing his bounty with others increases his odds of receiving meat when the tables inevitably turn.

Reciprocity

In general, a person is most likely to help another when the cost is small, the direct or indirect benefit is large, and the requestor can return the favor. The last part of this equation, reciprocity, is key. Don't believe me? Try this simple experiment: Dress in your Sunday best and head to your local bus stop. As the bus approaches, ask the other people waiting if anyone can make change for the too-large bill you have in your hand. Chances are excellent that several people will begin rummaging in pockets and purses. You may even have one or more people offer you some or all of the money needed to board the bus. Why? Because everybody who rides buses has been caught short from time to time, and the presence of the bill in your hand proves your ability to reciprocate. Someone

Tsze-Kung asked, saying, 'Is there one word which may serve as a rule of practice for all one's life?" The Master said, "Is not Reciprocity such a word? What you do not want done to yourself, do not do to others."
Confucius

> *We secure our friends not by accepting favors but by doing them.*
> Thucydides

who simply offers you money does so because the sum is small, and you are clearly in a position to return the favor at some future date (even though you both intellectually know that the odds of seeing each other again are remote).

Now try the same experiment again (at a different bus stop) wearing the oldest, grungiest clothes you own, and with no paper money in hand. You'll probably miss several buses before finally amassing enough tiny offerings to afford the ride. Be sure to check your local ordinances before setting out, because you may even be cited or arrested for begging if you're not careful. Nothing has changed, except people's perception of your ability to return the favor you seek.

Having accomplished both tasks and seen the results, ask yourself: When did you feel the most comfortable about asking for the help? I'm going to guess that you were fine during the first experiment, but embarrassed and reluctant the second time. OK, let's be honest: I highly doubt you performed either experiment at my direction, but I think you've probably experienced situations analogous to both scenarios in your life. In these cases, you probably felt most comfortable asking for help when you felt most confident about your ability to repay the favor.

If looking like one is both able and willing to return a favor increases the odds of receiving favors, then all beggars need do is clean themselves up, dress sharp, and wave the local equivalent of a twenty-dollar bill around, right? Actually, it's not quite that simple...

The Best Policy?

> *The best measure of a man's honesty isn't his income tax return. It's the zero adjust on his bathroom scale.*
> Arthur C. Clarke

It makes little sense to work for one's self and return favors if all one has to do is find ways to abscond with other people's resources. Life is a competition, and those who fail to safeguard themselves are on the losing end. It really is that simple, at least in theory. The only problem is that no society can long survive if all of its members are busy getting over on everyone else. Here again, the social contract comes into play where most individuals eschew certain actions in order to promote general peace and well-being. Even so, the temptation to fib a little from time to time is too powerful to ignore. I doubt that very many people can truthfully say that they've never fibbed, told a white lie, or acted other-than-honestly in order to gain some advantage or minimize some loss. Deception is efficient—so efficient as to be a predictable result of natural selection.

Leave it to evolution to guard against large-scale dishonesty. As we saw in Chapter 8, people are very capable at spotting cheaters. We routinely give favors to those who reciprocate and withhold favors from those who don't. Once cheated, we have a hard time restoring trust. Possession is nine tenths of the law for good reason.

> *Honesty is probably the sexiest thing a man can give to a woman.*
> Debra Messing

Needless to say, we are far more likely to trust friends and family than strangers. The closer the bond, the less likely we are to keep track of who is contributing what to the relationship. Go to lunch with coworkers, and you'll all probably invest several long minutes poring over the bill to make sure everyone contributes her or his fair share (if you haven't solved the problem by getting separate checks). Go to lunch with a friend, and you'll probably either split the bill (despite any cost differences between the meals), or one of you will treat. Go to lunch with a family member or significant other, and one or the other of you will probably pick up the tab without a second thought. At a certain point, you'll just stop keeping score beyond a subconscious sense that things either are or aren't in balance.

The flip side is that we are also far more willing to punish those we are closest to when they wrong us. Nothing spurs a detailed accounting like two friends, lovers, or spouses coming to a less-than-amicable parting of the ways.

Mental Illness

The latest version of the *Diagnostic and Statistical Manual of Mental Disorders* (DSM-IV-TR) lists 297 mental disorders in many categories, such as mental retardation, dementia, and disorders around cognition, substances (drugs), anxiety, schizophrenia, psychosis, mood, sleep, adjustment, and personality. I mentioned in the Introduction that the National Institute for Mental Health estimates that over 25% of the American population experiences a diagnosable mental disorder in a given year, and also listed the large numbers of people using

> *The statistics on sanity are that one out of every four Americans is suffering from some form of mental illness. Think of your three best friends. If they're okay, then it's you.*
> Rita Mae Brown

psychotherapeutic drugs. It is easy to conclude that humans are one sick species, this in a world that swiftly eliminates the weak and ill.

Are humans the one exception to nature's rules thanks to our highly developed social contract and resulting concept of human rights? Have psychologists gone overboard, finding disorders where none actually exist? Is there some middle ground between the two extremes?

The brain is an organ just like any other, and is therefore subject to physical trauma, chemical imbalances, etc. just like any other organ. Further, we've already seen how emotional programming creates chemical addictions that force us to repeat the same old patterns time and time again. Chapter 8 discusses the bell curve, and how the concept of normality in a given trait describes a certain range instead of an absolute value. Values outside the normal range could be classified as disorders.

In *The Enlightened Savage*, I made the case that much of what we call mental illness stems from our earliest mental programming and the subsequent layers of experiences and pattern repetition. In other words, what seems wrong with us may be happening because our brains are perfectly healthy.

The question of mental health may ultimately be one of semantics. A healthy response to a bad situation can cause plenty of ongoing suffering, and should be addressed by competent intervention where needed. Does this constitute mental illness? In the end, I think I like the perspective offered by my partner, psychiatrist Jennifer Cummings: Coughing is a perfectly appropriate response to dust, while chronic asthma is a different proposition, no matter the original cause.

I used to think that the brain was the most wonderful organ in my body. Then I realized who was telling me this.
Emo Phillips

126 | *The Natural Savage*
Discovering the Human Animal

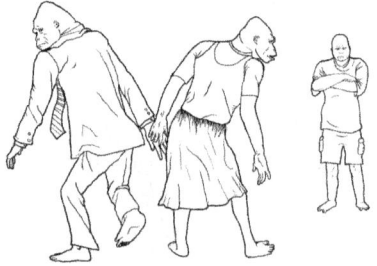

Chapter 11

It's Good to be King

> *Do not fall prey to the false belief that mastery and domination are synonymous with manliness.*
> Kent Nerburn

The amazing diversity of human cultures across the globe belies the fundamental sameness of people everywhere. All cultures have norms and rites around daily life and life events such as birth, puberty, marriage, possessions, laws, death, and status. Most if not all mammals experience some hierarchy and dominance struggles, and humans are certainly no exception. We use terms including respect, authority, deference, leadership, and elders to describe dominant individuals. Scientists studying wild animals are finding themselves forced to use the same terms when trying to fit observed behaviors into dominance theories. These terms imply that status has emotional components. Humans fear, admire, respect, and aspire to leadership. Why not animals?

Status in hunter-gatherer groups is often based on kinship. Children of high-ranking individuals tended to inherit their parents' stations in life, and the dynastic system of succession was born. This system is alive and well today.

Inherent Inequality

All hierarchical societies have leaders and followers, thus setting up a fundamental inequality. Depending on the species or the culture, high status comes with benefits such as preferential access to food and other resources, enhanced reproductive ability, money, and the power to command and direct the lives and deaths of others. Low status means less of everything.

Haves & Have-Nots

I cannot think of a single modern culture that does not include some sort of class system from beggars to magnates. The disparity between rich and poor can be staggering. A professional sports player may earn a salary 200 times larger than a working middle-class professional earning $100,000 per year. In 2004, the CEO of Colgate-Palmolive received $148 million ($148,000,000) in total compensation, or 1,480 times the salary of the aforementioned middle-class professional. As of July, 2007, the world's richest person had a fortune worth around $63 billion (63,000,000,000). The professional earning $100,000 per year would have to toil for 630,000 years to earn that much money—assuming he spent none of it.

You don't understand. I could have had class. I could have been a contender.
Budd Schulberg

The disparity between haves and have-nots gets even wider when we look at the lower socioeconomic classes. In the United States, the 2007 Federal minimum wage was $5.85 per hour, or $12,168 per year (assuming fifty-two 40-hour work weeks). Someone earning this pay rate would have to toil for over 8 years to earn $100,000 and for 5,177,514 years to earn $63 billion. California and Massachusetts fared better: In 2008, their minimum wages rose to $8.00 per hour or $16,640 per year. These fortunate folks need only work for 6 years to earn $100,000 or 3,786,057 years to earn $63 billion.

People at the lower end of the socioeconomic ladder are the ones most likely to perform physical labor under all types of conditions. They are arguably the ones in need of the best nutrition, health care, and other resources to keep them fit for duty. Sadly, the exact opposite is what happens. These people are often relegated to eating cheap food, whose nutritional value has been sucked out of it by excessive processing and replaced with exorbitant amounts of sweetness, fat, and salt to appeal to our prehistoric palates (see Chapters 15-17). They arguably need safe comfortable housing to rest after long workdays, top-notch childcare, and more. I would go so far as to say that they need these resources far more than someone like myself, who earns a living sitting in front of a screen.

The classes that wash most are those that work least.
G.K. Chesterton

Sadly, the exact opposite occurs: Those who most need the resources are often the ones most lacking.

The good news is that most societies with class systems allow at least some mobility. A rich person may find himself out on the street after a single disastrous day on the stock market. A poor person may win the lottery or manage to climb to a higher socioeconomic level by obtaining the education and skills needed to thrive in a new career. This isn't as easy as it sounds; societies have mechanisms in place that thwart social mobility—and upward mobility in particular—with varying degrees of effectiveness.

Hunter-gatherer societies certainly have varying degrees of status and confer increased benefits on their higher-ups. The key difference is that an observer will be hard-pressed to find the same disparities between rich and poor in these supposedly primitive societies. The key may be agriculture, specifically the lack of agriculture. It's hard to starve someone or keep them in any kind of cycle of dependency when they are perfectly capable of catching or foraging their own food and building their own shelter. Modern societies have created specialists who control the production of—and access to—food and shelter. Combine this control with humanity's inherent status struggles, and the class system is born.

Castes

There is always more misery among the lower classes than there is humanity in the higher.
Victor Hugo

The socioeconomic class system has its drawbacks, but generally does allow some mobility. Some societies, such as India, took class one step further by creating *castes*. Insect societies (bees, ants, termites, etc.) breed individuals to perform very specific tasks such as foraging, reproduction, and protection. Insects cannot change caste; they perform the same function their entire lives. Human castes are similar: Children are born into specific castes and marry others of the same caste. Castes are limited to people with identical or nearly identical statuses, ranks, and careers.

Indian society consists of the following five castes:

- *Brahmans* are intellectual and spiritual leaders, analogous to intelligentsia and priests.

- *Kshatriyas* are the warriors and rulers. Soldiers and politicians fall into this caste.

- *Vaishyas* own land and conduct agriculture and commerce.

- *Shudras* are laborers (skilled and unskilled) and servants.

- *Untouchables* carry out the most menial jobs under the worst conditions. I once saw a picture of a man emerging from the sewer he had just cleaned wearing nothing but a loincloth.

Each of these castes contains divisions and subdivisions by specialty, location, etc. that are used for (among other things) arranging marriages. Castes are therefore *endogamous*, meaning that people are typically expected to marry others from the same caste, thus ensuring that offspring will belong to the same caste from birth.

The caste system in India is starting to loosen up. Southern India tends to be more liberal than the northern regions, and cities tend to be more liberal than villages that may not have much outside contact. Some might say that the loosening up has quite a way to go. One Indian friend of mine was the only woman in a class of 600 women to be allowed to come to the United States and start a career without being married first. Women marrying arranged males of their caste in their early 20s is still the norm in India.

Dynasties

Animal leaders are not secure in their positions. Old age, injury, illness, or a strong enough challenge by one or more rivals can knock them from the top spot at any time. Aspiring chimpanzee leaders form alliances designed to help them take power, the accomplices participating with the promise of preferred treatment upon staging a successful coup. There is also no guarantee that a leader's offspring will ever assume the mantle of power.

Human leaders are not inherently any more secure in their positions. A society with class mobility means that challenges can come from any quarter. The caste system normally ensures that challenges will only come from within one's own caste, but does little else to reduce the threat to power beyond

We inhereit from our ancestors gifts so often taken for granted... Each of us contains within... this inheritance of soul. We are links between the ages, containing past and present expectations, sacred memories and future promise.
Edward Sellner

reducing the number of potential challengers. Here again, there is no guarantee that any leader's offspring will enjoy the same level of power.

Enter the *dynasty* system that guarantees succession of leadership by family ties. All a leader need do is produce a (generally male) heir for power to remain in the family. Power generally goes to the firstborn—or at least the firstborn male. Things can get more than a little sticky if no heir is produced or if only females are born. For example, Henry the VIII barred his daughter Elizabeth from the throne of England in favor of his younger son Edward VI. Edward willed his fourth cousin Lady Jane Grey to the throne, but this was overturned in favor of Henry's sister Mary, and then Elizabeth, following Mary's death. Failing to produce an heir, Elizabeth willed the throne to James I, thereby ending the Tudor dynasty and ushering in the Stuart dynasty that lasted until 1707. Ancient Egypt went through thirty-one dynasties, not including the Macedonian and Ptolemaic times. China had over a dozen dynasties.

These are just a few examples, but the overall lesson is clear: Leaders of any species will do whatever they can to hold onto power and pass it on to whom they choose. Chimpanzees and bonobos haven't made the mental connection between copulation and offspring. Given this awareness, I can only speculate that some form of dynasty system would soon follow.

Alpha

The first duty of a leader is to make himself be loved without courting love. To be loved without 'playing up' to anyone, even to himself.
Andre Malraux

Alpha is both the first letter in the Greek alphabet and the term used to describe those animals or people holding the highest positions of power in a given group. It really is good to be the king. High-ranking animals are well groomed and well fed. They tend to be taller than lower-ranked individuals, stand tall, and carry themselves with poise and confidence. They command respect from all below them, and never hesitate to use mental or physical force when necessary to foil or quell challengers. They tend to be bigger and stronger than others, the better to overpower rivals.

This is not to say that leaders get a free ride. With great power comes great responsibility, and subordinates will not long tolerate a leader who shirks her or his duties. A leader must constantly be on the lookout for anyone who would challenge the status quo. She or he must reward loyal followers, some of whom may have helped in the rise to power, and must place the group above any individual—even family. Failure to do this violates the reciprocity principle we discussed in Chapter 10, and can have dire consequences.

Peacekeeping is another vital leadership role. A leader must either directly or indirectly keep the peace among subordinates. Leaders must be able to serve as mediators or riot police when needed. An angry population is an unruly population whose members may vent their frustrations on the leader. More than one human ruler has been toppled by popular revolt.

Peacekeeping also extends to protecting the weakest members of the group or society from undue harassment and persecution. An *omega* may be the group's whipping boy, but there must be limits. Killing off or otherwise driving out the omega means another must assume that lowly position. It is to everyone's benefit to keep the lower echelons around by making sure they are least moderately content with their lot. Rebellious masses can really ruin a leader's day.

A group can only function as a group to the extent that it shares a common agenda. Leaders set the group's agenda and make sure it gets carried out. The group must also be kept at least reasonably intact against external threats, such as predators or rivals. And, where possible, the group should take advantage of its strength to take resources from weaker groups.

It is indeed good to be the king in many ways. In other ways, the position may leave a little to be desired.

War is a cowardly escape from the problems of peace.
Thomas Mann

Omega

Omega is both the last letter in the Greek alphabet and the term used to describe those animals or people holding the lowest positions of power in a given group. The bottom of the social hierarchy or *pecking order* is just that: Any chicken in a flock can peck the bottom-ranked chicken, who cannot peck anyone. Mid-ranked chickens may be pecked by higher-ranking chickens, but may also peck chickens of lower rank. A

The trouble with being poor is that it takes up all of your time.
Willem de Kooning

chicken's place in the pecking order also determines who gets to eat first, and who must wait. Among wolves, the alpha male and female enjoy the first and best portions of meat from kills, while the omega must scrounge or even beg for scraps.

> *The petty economies of the rich are just as amazing as the silly extravagances of the poor.*
> William Feather

A wolf pack may have one omega, but other animal and human groups may have multiple omegas. For example, there are over 100 million "untouchables" in India. Some 36 million Americans live below the Federal poverty line. Worldwide, about 2.75 billion people live on less than two dollars per day, and about 50,000 people per day die of poverty-related causes. Visit www.poverty.com, and you'll see a map with children's names and photos popping up every second or so to mark yet another death. According to that site, about seven children died in the time it took me to write this very sentence. The key difference between animal and human omegas is that even the lowest-ranking wolf or chimpanzee gets to eat. Human omegas have no such guarantee.

I've already touched upon the vast disparity between human haves and have-nots. The *Gini coefficient* measures the inequality of income distribution. Lower numbers mean a smaller disparity and vice-versa. The *CIA World Factbook* dated December, 2006, credits Albania with the lowest Gini coefficient (26.7) and Zimbabwe with the highest (56.8). The United States has a Gini coefficient of 45. There are only seven nations listed with higher numbers.

> *Of all the preposterous assumptions of humanity over humanity, nothing exceeds most of the criticisms made on the habits of the poor by the well-housed, well-warmed, and well-fed.*
> Herman Melville

In general, lower-echelon group members serve as the group's muscles and scapegoats in both animal and human societies. They are less-well groomed and have restricted access to both feeding and reproduction. In general, low status means a lesser chance of passing on one's genes, which is in perfect accord with the concept of survival of the fittest. Humans are an interesting exception to this general rule. People in lower socioeconomic classes tend to have more children than those in higher societal tiers. In late 2007, Joseph Ratzinger (Pope Benedict XVI) chastised Europeans for low birth rates. Not coincidentally, Europeans have some of the highest standards of living on Earth, even higher than the United States. Meanwhile, the world's poor are reproducing at alarming rates. The primary reason for this is that humans don't restrict reproductive rights by status or rank to a sufficient degree to limit population growth. Freed of this restriction, there are several valid reasons why reproducing early and often makes sense for those in lower socioeconomic classes. Chapter 20 discusses this seeming paradox in more detail.

Submissive Behaviors

Lower-ranking humans and animals have specific behaviors that indicate their submissive status. These behaviors serve to pacify higher-ups by sending the clear message that no challenge is forthcoming; the lower-ranked individual knows her or his place, and doesn't want trouble. A few examples of submission behavior include:

The marvel of all history is the patience with which men and women submit to burdens unnecessarily laid upon them by their governments.
William H. Borah

- *The handshake.* Handshaking among humans is usually traced back to Romans, who clasped forearms to ensure that nobody was carrying hidden knives. This in turn traces its way back to chimpanzees, where the lower-ranking individual extends a limp hand. Even today, the strength of a handshake is a strong indicator of confidence and perceived rank.

- *Appearing smaller.* Submissive animals may tuck their tails between their legs and crouch or grovel, thereby making themselves appear smaller to their superiors. Enlarging one's body (such as by dominant or attacking chimpanzees or by hikers attempting to scare off a bear) is a direct challenge that is recognized by many species (which is why spreading arms and legs wide is a good way to convince a bear to leave you alone). Other examples include bowing and curtseying, and even prostrating one's self before a superior.

- *Apparent vulnerability.* There's a reason company executives sit behind massive desks. Those huge masses of wood form impregnable shields between the superior and his minions, who must often sit with their backs to a door—a distinctly uncomfortable place to be. It's hard to attack someone when they're already exposed and weak.

- *Submissive vocalizations.* From the pant-grunts of submissive chimpanzees to the respectful and even reverent or ceremonial words humans use in the presence of leadership, sound plays a vital role in conveying status. In conversation, a lower-ranked person will mimic the pace and tone of the other's voice. Some of this may not be apparent to the casual listener or even to someone listening intently for these subtle clues. Nevertheless, those clues are both present and heeded. They are one of the many ways we form impressions about other people. Try this experiment sometime: When someone is talking rapidly and excitedly, respond in a calm soft tone. Keep this up no matter how the other person reacts. You'll soon find them following suit. We communicate our status every time we speak. Your getting the other person to modify their voice communicates your higher status.

> *To have respect for ourselves guides our morals; and to have a deference for others governs our manners.*
> Lawrence Sterne

- *Deference to authority.* I've stopped at my fair share of car accidents. No scene was complete without a knot of bystanders milling about aimlessly, staring, whispering to each other, or perhaps making some feeble attempt at rendering aid. Within moments, I'd have people helping stop traffic, searching for other victims if necessary, calling the fire/police/ambulance departments, and helping out in any other way needed to secure the scene and begin helping the victims. All I had to was point at someone, tell them what to do in a clear confident voice, and they'd comply. Can you believe that nobody has ever asked me who I am or what my qualifications are for taking charge? I happen to have served in the US Coast Guard Reserve, been a volunteer firefighter, emergency medical technician, and CPR instructor. The people scurrying to obey me had no way of knowing this without asking—which they never did. The rule is simple: Act like you're in charge and know what you're doing, and people will more often than not follow you without question.

- *Averting eyes.* Staring is a direct challenge recognized by both humans and animals. Stare at someone long enough, and they'll either look away (a sign of submission to your authority) or ask you just what you think you're looking at (challenging your challenge). I don't recommend trying this one as any sort of experiment.

- *Closed-off body posture.* People who are nervous tend to close themselves off. This can be as subtle as clasping one's hands in front of one's self, or can be a full self-embrace. Confident people have no problem with open postures that leave the body vulnerable, because they are confident that no harm will befall them. Confidence is a hallmark of high rank.

- *Smiling.* Smiling is an appeasement expression that reduces tension in humans and apes.

This list is by no means exhaustive, but it should give you a good idea of what to look for when judging a person's rank. Pay special attention to how you act around different people and how your voice either leads or follows conversations. You'll soon get a good idea of how you see yourself in the pecking order and how others see you.

Apes are far better at detecting a person's mood and status, because they pick up on the many subtle clues our body language gives off. They don't rely on spoken words, and therefore don't have any contradicting information coming between them and their impressions. Human status detection may often rely on subconscious impressions and clues. Don't let this comparison fool you, though: Humans have a deep-seated need to know their place in the hierarchy, and are constantly transmitting and receiving statues cues. Chaos would reign without those cues.

An outsider watching a group going about its business can quickly tell who's where on the totem pole. A good salesman visits clients many times in order to carefully study relationships, centers of power, and influence, while paying special attention to interpersonal connections in order to find the best route of entry to make the sale.

Jockeying for Position

Nothing lasts forever. Every leader must eventually face the end if her or his reign through aging, illness, injury, loss of favor, or upstart challengers and their allies.

Stable vs. Unstable Hierarchies

Want to get something done quickly and efficiently? Establish a clear hierarchy with a strong chain of command. The lower-ranking folks may not appreciate it too much, but then again no one will be able to argue with the productivity. This system

> *In accordance with our principles of free enterprise and healthy competition, I'm going to ask you two to fight to the death for it.*
> Monty Python

may allow individuals to shuffle up and down the ladder, but the ladder itself will be stable, and everyone will know her or his place. It may be possible to limit the spread between the top and bottom rungs, but humans are so wired for hierarchy that any attempt to eliminate it will backfire. Ants have an extremely rigid hierarchy rooted in a biological caste system and you won't find a more efficient workforce.

One good way to stabilize a hierarchy is to help the winners. Monkeys do this, and enjoy relative peace and quiet. Chimpanzees switch allegiance between winners and losers, resulting in very unstable hierarchies. Stabilizing a human hierarchy is possible—to a point. People respond to community needs, but not at their own expense, a point that Karl Marx failed to grasp.

Unstable hierarchies cause a lot of stress, and may therefore benefit from a convenient individual (scapegoat) on which to redirect blame and promote unity with little fear of reprisal. Scapegoats tend to be both targets and innocent of the charges leveled against them.

Politics

Politics is supposed to be the second oldest profession. I have come to realize that it bears a very close resemblance to the first.
Ronald Reagan

Anyone who has belonged to any company or organization for any length of time has probably encountered *politics*, the art of dealing with people by means of intrigue or strategy to obtain power or control. Niccolò Machiavelli is one of the most famous political thinkers in history. His most noted work, *Il Principe* (The Prince), describes how a ruler can maintain control over the realm. Some readers have mistakenly interpreted Machiavelli's philosophy as "ends justifying the means," but this is not the case. While he does not eschew the use of force or "evil" means, Machiavelli does place strict limits on their use. Still, the dictionary defines *Machiavellian* as "subtle or unscrupulous cunning, deception, expediency, or dishonesty." Does this bear an uncanny resemblance to modern politics, or is it just me?

I've worked in enough companies and belonged to enough organizations to know that there are often two power structures: the official and the unofficial. The official power structure follows the organization chart, with everyone given a job title and area of responsibility. The unofficial power structure may or may not follow the organization chart, and is built on friendships, alliances, and maneuvering for position. Two equally competent people in the same official position may have far different experiences based on their unofficial status and power within the group. The group grooms and promotes whoever is in vogue, regardless of performance or other merit, while passing over other candidates who don't belong to the "in" crowd. Gossip and rumors can run rampant, and someone who appears friendly may not be friendly at all. The whole system is built on *quid pro quo*, where the one helped to power attains a position to repay the favors. Finding out what a person will do once they attain rank can be as easy at looking at who is supporting that person and what's important to those people.

The people helping the one gaining power must be in a position to help, that is, they must have the resources and means to do so. This typically means that they must have some power, influence, or other resources of their own. Omegas have little chance of inclusion because they have the fewest resources available. Those in power may give some stirring speeches about helping the little guy, but few take any substantive steps to make any radical changes. The key is to maintain the status quo while keeping the low-ranking masses just happy enough to prevent an outright revolt. Revolts, when they occur, are typically led by people pledging to return power to the masses, only to establish a new elite while keeping the low-ranking people in much the same condition.

Democracy is an interesting experiment in which power is supposedly given to the masses, and where the importance of rank is thereby reduced. Leaders are called "public servants," because they supposedly serve the interests of the common person. The only problem with this idyllic theory is that only those potential leaders with the resources to mount a serious campaign are available for election. Who has the resources to help candidates win? You guessed it: the elite. Still, citizens in a democratic society are able to demand a certain level of treatment, which forms the foundation of a constitutional government.

Chimpanzee politics are similar in many ways. An aspiring leader forms a coalition of supporters who help him (chimpanzee societies are ruled by males) topple the current leader

---

Man is by nature a political animal.
Aristotle

---

in exchange for favors. Theirs is a simple and sometimes brutal system; on the other hand, it deserves points for its sheer simplicity. Bonobos also have politics, but theirs are far less violent than our chimpanzee cousins. Female bonobos pull the strings that help others gain rank.

Ladder Climbing

I've always said that in politics, your enemies can't hurt you, but your friends will kill you.
Ann Richards

Chimpanzees, bonobos, and humans begin learning how to curry favor, fit in, and jockey for position at a very early age. Fitting into and being part of the group are incredibly important, and status gained during childhood may better position one for adulthood. Children who achieve a certain level of acceptance and respect tend to be happy and confident, while those who fail tend to be depressed. A child (or anyone for that matter) who is unlucky enough to be expelled from any group s/he belongs to suffers a major disaster, because the deepest part of our prey brain knows that safety comes from numbers. As I said before, a lone human is a dead human.

Socioeconomic mobility and other forms of acceptance may be limited by one's class/caste or by any number of factors such as ethnicity, religion, where someone lives, etc. Still, each layer of the socioeconomic cake has its own sub-layers that resemble the hundreds of divisions within the five main Indian castes, and children will strive to rise to the top of whatever heap they belong to. My son Logan is no exception. I often watch him and his friends play, fight, make up, and form and break alliances. The sight bears a too-strong-for-coincidence resemblance to similar behavior among apes and other mammals.

The lesson is clear: Status is extremely important to humans and animals alike. Authors Steven J. C. Gaulin and Donald H. McBurney define *self-esteem* as self-evaluation of one's place on the totem pole. High status is a source of pride (raised self-esteem) and elevated testosterone levels that encourages further efforts following a victory. Failure encourages withdrawal and triggers shame (lower self-esteem).

Rank theory postulates that depression evolved as a response to loss of rank. Depression and its resulting apathy ease the pain of losing and helps people cope with defeat. This prevents the loser from suffering further injury, while also preserving the stability and efficiency of the group by reducing aggressiveness and preserving the pecking order.

Roaring Mice

An interesting phenomenon sometimes allows a small group to wield inordinate power when placed in between two larger, evenly matched groups. This happens because the small group becomes the "swing vote" that can tip outcomes in favor of either large group. In this case, the large parties will woo and defer to the small groups.

An elephant: A mouse built to government specifications.
Robert Heinlein

Friends in Low Places

Remember that there are two types of hierarchy: the official and the unofficial. Want to get a meeting with the CEO? You'd better be on good terms with the secretary! There are entire training programs dedicated to befriending people in low places who hold the keys to accessing power.

Ritualized Combat

Conflict among members of a society is inevitable, but peace is also crucial, because violence and killing are not good for the species. A successful society must therefore include established methods of resolving conflict and making up afterward. Ritual fighting allows combatants to vent their anger and resolve the conflict, leaving real fighting as the last resort. The defeated enemy poses no further threat, and can be ignored. An enemy who can signal defeat can both avoid further harm and appease the winner.

Ritual is the way you carry the presence of the sacred. Ritual is the spark that must not go out.
Christina Baldwin

Men tend to take risks and hide real or perceived weakness far more regularly than females, lest any sign of weakness be exploited at an inopportune moment. We are also quick to argue about most anything, whether or not we care about the subject. On the other hand, we tend to resolve conflict more quickly, and with far less lasting rancor than women. Boys get into visible conflicts more often than girls, but soon forget what the fuss was about. Beneath their calm exteriors, girls report experiencing just as much conflict as the boys, and remembering those conflicts far longer.

Why do we fight? Animals fight for status and to defend or take over territory. Among humans, men commit most of the violence, most often for reasons involving status or honor.

Religion

> *My religion consists of a humble admiration of the illimitable superior spirit who reveals himself in the slight details we are able to perceive with our frail and feeble mind.*
> Albert Einstein

The concept of a god may have stemmed from the idea of a hidden "über-alpha" that guides forces beyond human control. Alpha humans guide all things that the group can control, so why not ascribe all things beyond that control to an unseen super leader?

Create a caste or class of specialists (priests) who claim to have some direct connection to this invisible chieftain, and who claim to know and speak divine wisdom and commands, and religion is born. This priest class can gain power beyond their ordinary means by inventing complex rituals and prescribing punishments for displeasing the god(s). This power can even sway ordinary kings, dictators, presidents, CEOs, and other leaders' decisions and policies—not to mention the core beliefs of the masses themselves. Priests, imams, rabbis, ministers, monks, and other religious professionals comprise a tiny fraction of the overall population, but wield enormous power.

Religion has played a key role in most—if not all—of history's most significant events, and is single-handedly responsible for more death, destruction, oppression, and general misery than all other causes combined. I believe this is because humans fear death, and the promise of life eternal (if one follows certain rules and obeys the priests without question) is an extremely potent motivator.

> *All formal dogmatic religions are fallacious and must never be accepted by self-respecting persons as final.*
> Hypatia of Alexandria

You may have correctly guessed that I have huge problems with organized religion; however, this does not mean that I reject spirituality. As I said in Chapter 1, I don't think for a moment that evolution and creation are mutually exclusive concepts, nor does my disdain for religion mean that I can't accept the idea of an immortal soul. I explain these beliefs and the reasons behind them in *The Divine Savage*.

King for a Day

The news isn't all bad for people in the lower tiers of any given human society. The tremendous variety of companies, clubs, circles of friends, and the Internet hold out the hope that just about anyone can achieve some level of leadership. Some of humanity's unique inventions even allow people to fake power.

Tiny Fiefdoms

In *The Human Zoo*, author Desmond Morris puts forth the concept of a *super tribe*, a group of humans that is so large that nobody can possibly know everyone else. Large companies, towns, cities, states, and nations are examples of super tribes of people living and working in groups that are both many thousands of times larger and more densely packed than the hunter gatherer tribes we evolved from.

> *Call it a clan, call it a network, call it a tribe, call it a family. Whatever you call it, whoever you are, you need one.*
> Jane Howard

Clubs, companies (or individual departments, offices, etc.), groups of friends, and more are examples of *pseudo tribes* or subsets of the larger super tribe. Pseudo tribes offer plenty of opportunities for everyone to have some leadership and power. Put any group of two or more people together, and a hierarchy will immediately emerge. Even the most balanced relationship between two people has an aspect of power to it that may be either fixed on one person or shared between the two.

Becoming king may be impossible, but being the president of your local Toastmasters club or a member of your town council is an easy undertaking. Merely establishing yourself as an expert on a certain topic can give one a degree of power and control. And, if all else fails and you become homeless, you can always stroll slowly across a busy street. I don't say this sarcastically; on the contrary, this is a prime example of how the human race offers an unprecedented number of leadership opportunities. One can even be part of multiple hierarchies and enjoy several positions of leadership if one so chooses.

It is important to remember that leaders at all levels share the same responsibilities I noted above. The wide menu of available options may make it easy to obtain some measure of control, but the only way to retain control is to lead wisely.

Faking It

The secret of success is sincerity. Once you can fake that you've got it made.
Jean Giraudoux

The human concepts of class mobility, money, and credit have given rise to a phenomenon that is unique in the animal kingdom: faking dominance. Anyone with enough money or a credit card can buy the power suit, fancy car, and other trappings of power and leadership. One can even purchase courses that purport to teach leadership skills and behavior. There is even a whole industry devoted to short-circuiting the human mate selection process in order to obtain sex.

Faking power is a risky undertaking. Remember that humans excel at detecting cheaters and impostors, and reserve a special contempt for those who game the system. The consequences of being exposed as a fraud should make anyone think twice about making such an attempt. Even if the people one associates with don't catch on, chances are decent that the cheater is expending resources at an unsustainable rate, meaning that her or his apparent rise in status will be short-lived. The general public might not catch on, but the credit card companies most certainly will!

I am not saying that someone should simply roll over and accept her or his lot in life. Aspiring to power and status is one of the strongest instincts humans have. I am saying that one should temper this instinct and look for genuine opportunities to gain status. Start as small as you have to, demonstrate competence and confidence at that level, then work your way up. This may take a lot longer, but the results will also last much longer.

144 | *The Natural Savage*
Discovering the Human Animal

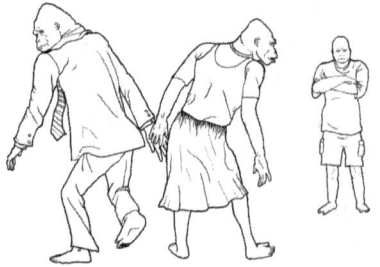

Chapter 12

War & Conquest

> *War is a series of catastrophes that results in a victory.*
> Georges Clemenceau

Conflict within societies and even families is unavoidable, as individuals walk the fine line between serving their own interests and those of the group, while balancing short-term opportunities against potential long-term gain. Societies must therefore tolerate and manage a certain level of internal strife by defining both acceptable limits and mechanisms for restoring the peace after a conflict. Our individual interests may conflict with the group's, but we still need the group for protection and status, and this requires a certain degree of acceptance and tolerance.

No such problem exists with outsiders, who provide neither support nor protection, and who are competing for the same limited resources. Humans are highly territorial and will actively defend that territory, be it land, physical possessions, or intangible assets (such as copyrights). There is little reason to trust or even like outsiders, except where significant migrations occur on a regular basis, such as female bonobos who leave their birth group to join another one. Here, violence against another group risks killing off individuals who are carrying one's own DNA, and is therefore not a good idea.

Us vs. Them

Nothing brings a group together like a perceived opportunity or threat from outside. Groups have numerous ways of portraying themselves as superior to all others. Modern examples include flags, anthems, and oaths of allegiance. A person who migrates from one society to another must usually undergo an indoctrination process ending in a pledge of loyalty to the new society. Most nations restrict at least some jobs to citizens and other jobs to born citizens only. For example, one can only be President of the United States if one is an American citizen by birth.

If members of a group see themselves as superior, that can only mean that they view outsiders as inferior. From there, it is only a small step to adopt two different forms of aggression, one for intra-group conflict, the other for conflict with other groups. The former will always include restraint and limits that do not apply to the latter. In general, we will do all we can to cooperate with members of our own group, because we depend on them for survival.

Warfare probably has its roots in occasional clashes that occurred when neighboring groups encountered each other in a manner similar to that of today's chimpanzees. The development of agriculture and resulting transition from hunter-gatherer to settled societies with increasing specialization of professions enabled the creation of standing armies that made genuine warfare possible. hunter-gatherer groups were (and remain) perfectly capable of conducting organized combat; the primary differences between these groups and "civilized" societies are those of scale, intensity, destructive potential, and complexity. A short, small-scale conflict is easy to start and manage. A long, large-scale war poses a complex logistical challenge.

There are few if any hard and fast rules about human inter-group relations. The inherent hostility of these relations is tempered with a strong desire for peace. Nations negotiating, treating with, making war on, and allying with each other are very similar to individuals jockeying for position within a single group. The self-interest of each nation bumps up against the social contract, which reminds us that it may be better not to fight.

Insanity in individuals is something rare but in groups, parties, nations and epochs, it is the rule.
Friedrich Nietzsche

Did you know that every two hours the nations of this world spend as much on armaments as they spend on the children of this world every year?
Peter Ustinov

Violent Nature

> *Democracy don't rule the world, You'd better get that in your head. This world is ruled by violence, but I guess that's better left unsaid.*
> Bob Dylan

Nature is violent. Life itself requires death in order to continue. Bacteria decompose dead plant and animal tissues. Plants live on the composted remains and are killed by hungry animals, who are in turn killed by predators. There is nothing whatsoever humane about the way predators hunt, chase, and kill their prey. On the contrary, a quick stunning followed by a rapid death in a slaughterhouse seems far more humane than the animal running for its life only to be tackled and strangled/shaken/etc. to death in a process that can take well over a minute. Predators eventually die, and the whole cycle begins anew. This example is an overly simplistic view of the food chain, but it does illustrate my point. You are alive because numerous plants and animals have died. When you die, nature will (in theory) decompose your mortal remains, and you too will become another spoke in the great wheel of life. I say in theory, because embalming and burial techniques effectively stop the natural recycling process in its tracks.

Conflict begins in the womb, a tug-of-war between fetus and mother looking out for their own interests. The fetus wants to live; the mother is evaluating the fetus's fitness as a carrier for her genes, and will spontaneously abort (miscarry) the pregnancy if the fetus is found wanting. There is nothing nonviolent about being killed and ejected from the womb. This conflict continues after the baby is born, taking on such forms as sibling rivalry and attempting to climb the social ladder. It may even manifest itself in ritualized or even actual fighting. The reason is simple: With high status comes improved reproduction potential. High reproduction potential increases one's odds of passing on the all-important genetic material to the next generation.

> *Victory attained by violence is tantamount to a defeat, for it is momentary.*
> Mahatma Gandhi

Conflict and violence are two of the cornerstones of life. I will go so far as to say that outsiders (people who don't belong to the group/family/tribe/society) provide a much-needed outlet for pent-up aggressions. The further removed a person is from a given group, the easier it is for the group to attack that person. As I type this, millions of American people are peacefully going about our lives while American soldiers are waging bloody wars in Iraq and Afghanistan. We may harbor varying degrees of concern over the lives and resources being spent on these wars, but the fact remains that most of us are far removed from the heat of battle.

If conflict is inevitable, and if outsiders provide sinks for pent-up aggression, then societies may turn on themselves in the absence of external enemies. For example, African Americans experienced over a century of various forms of oppression from slavery to various forms of apartheid that only began crumbling in the late 1960s. Today, the African American community remains plagued by violence. The difference is that this violence is far more likely to come from within the community—and target others within the community—than from outside sources. Rival gang warfare in Los Angeles, California is just one example.

Violence may have a beneficial effect on the species: Competition for scare resources can reduce a population to the local carrying capacity, with enough resources available for the survivors. These survivors will also be the ones to reproduce, arguably producing stronger, more resilient offspring than would otherwise be possible. That said, large-scale violence among social animals is a relative rarity, and no species except humans has the capability to render itself extinct while taking the rest of the planet with it.

Animals at War

Nonsocial animal species may fight and kill each other on an individual basis, but *wars* (groups of animals fighting each other) are impossible since the animals are loners by nature. Social animals are inherently capable of warfare, and various species can and do fight wars. Ants and termites have dedicated soldier casts, and their battlefields can resemble the most bitter human struggles, albeit on a small scale. Chimpanzees fight brutal wars complete with scouts, sentries, weapons such as rocks and sticks, and raiding parties.

Bonobos have never been observed to fight wars. As primatologist Frans de Waal states, chimpanzees resolve sex issues with power, while bonobos resolve power issues with sex. This makes sense, because bonobo females migrate away from their birth groups to join other groups, thus avoiding inbreeding. Attacking a neighboring bonobo group therefore carries

From the moment of birth, when the stone-age baby confronts the twentieth-century mother, the baby is subjected to these forces of violence, called love, as its mother and father have been, and their parents and their parents before them. These forces are mainly concerned with destroying most of its potentialities. This enterprise is on the whole successful.
R. D. Laing

Chapter 12
War & Conquest

> *Violence isn't always evil. What's evil is the infatuation with violence.*
> Jim Morrison

the unacceptably high risk of harming one's own genetic material, since it's very likely that at least some members of the other group are blood relatives. Besides, bonobo females are outright floozies compared to chimpanzees, meaning that males have no reason to fight for mates. Even better, bonobos resolve power issues with sex, as de Waal notes. I can't speak for anyone but myself when I say that having sex is far preferable to fighting. I can say that most if not all bonobos would probably agree with me.

Humans exhibit both chimpanzee and bonobo behavior between our own groups/societies. Our wars are much, much worse than those of any other animal species, because we alone possess the means to eliminate all life on Earth. On the contrary, our peaceful relations tend to be better and far more long lasting than those of our animal brethren.

Humans at War

It's more than a little interesting that nations will routinely jail and even execute a person for killing one other person, while at the same time pinning medals and heaping praises on people who have each killed many people and who have, in fact, gone out of their way to do so. We call the former *murderers*, *animals*, and *criminals* while reserving the terms *soldier*, *patriot*, and *hero* for the latter. Why is it acceptable to commit wholesale slaughter against members of a different society, while killing one member of the same society is a heinous offense?

> *Any war that requires the suspension of reason as a necessity for support is a bad war.*
> Norman Mailer

The short answer is that humans, like many social animals, have a powerful built-in inhibition against killing individual people, especially members of our own society. Remember that primordial societies were small, often less than 100 people, making it more than likely that members of a given group shared at least a little genetic material with everyone else. The entire point of living is to pass on genetic material. Killing someone from the same society risks harming one's own genetic material, and is therefore not a good idea. Modern human societies with their huge numbers make the odds of killing a blood relative extremely low, but they are a very recent invention—too recent for us to have lost our built-in restraint. I once again point to Rousseau's social contract and the complex maintenance behavior I've already discussed as evidence.

I've already pointed out that humans are very territorial. Territory is essential for survival, because we live off the land either directly or indirectly, and therefore require enough *lebensraum* (living area) to provide for ourselves. Outsiders threaten territory and thus survivability and must be kept out. On the other hand, expanding territory through conquest can be a perfect way to increase group resources (and therefore vitality) while venting internal strife at the same time. A society provides protection for its members against both predators and outsiders regardless of family ties. Thus, people place a higher value on the lives of others in their societies than they do on the lives of outsiders. This is how people can mourn the loss of a friend or relative while reading about thousands of war dead in a distant land over their morning meal without any emotional reaction.

Humans tend to view ourselves as superior to animals, conveniently forgetting that we are animals ourselves. This has all sorts of interesting consequences ranging from overly anthropocentric scientific models to the tendency of groups in combat to dehumanize their enemies. We refer to our enemies as *japs* (Japanese in World War 2), *huns* (Germans in World War 1), *jerries* (Germans in World War 2), *gooks/dinks* (Vietnam), *skinnies* (Somalia), *mujjes* (Iraq/Afghanistan), and many other even more colorful terms. Our vocabularies are rife with ethnic slurs such as *nigger, kike, wop, spic, mick, kraut, chink, greaser,* and *honky*. We also use such generic terms as *animal, heathen,* and *savage*. Killing a fellow human being may be a tall order. Killing a gook is all too easy as is eliminating an entire population of savages.

An invasion of armies can be resisted, but not an idea whose time has come.
Victor Hugo

There is some speculation that warfare is an outgrowth of group hunting, which requires cohesion and strong bonds among the hunters in order to be successful, particularly against large prey such as mammoths. This group cohesion and bonding forms the basis for the military chain of command and organized warfare, according to this theory. I find this idea plausible in some respects; however, chimpanzees and other animals wage war well enough without any hunting-induced efficiencies. Also, highly social animals (humans included) already have hierarchies that can be adapted for fighting purposes. Throw in the built-in bias against outsiders, and warfare has a solid foundation. I believe that the greatest contributions hunting makes to warfare involve strategies and tactics because killing humans is no different on its face than killing any other animal. The sequence of events described in Chapter 7 apply just as well to warfare as they do to predation.

Why Fight?

> *Be kind, for everyone you meet is fighting a hard battle.*
> Plato

My conversations with war veterans plus my own limited experience on active duty (I served during wartime but never saw combat) lead me to the conclusion that the reasons for going to war may be extremely complex, as we'll see in a moment. Once on the battlefield, soldiers fight to protect themselves and their own comrades. The distinction is clear: A person may volunteer for armed service because of some sense of duty to God and country. That same person actually pulls the trigger lest he and/or his buddies get killed. Most soldiers neither love their comrades nor hate their enemies, a fact poignantly illustrated by the Christmas Truce of World War 1, where warring English, French, and German troops temporarily halted the fighting and met in between their respective lines to exchange gifts, bury the dead, sing carols, and even play soccer.

War has very humble roots in the needs to defend one's self, family, and tribe/society. These roots derive in turn from the fundamental need to pass on the all-important genes. Defending against attackers is therefore necessary. As for offense, the desire to increase one's own status and access to resources is just as powerful as the need to defend against others bent on doing the same thing. Increasing the group's territory increases both survival and reproductive odds, while weeding out the group's weaker or less cunning (or simply less fortunate) members through casualties. As a side benefit, a person who fights well can reach the pinnacle of power and status within the group, as the ongoing litany of nations run by military veterans attests. Heroes and saviors receive much more favorable treatment than run of the mill combatants, who in turn tend to be held in higher regard than ordinary civilians. Were I to run for President of the United States, I guarantee that my lack of combat experience would be held against me, because I presumably lack the experience required to effectively command the armed forces. In fairness, there may be some merit to this argument.

Territory and resources are what motivate animals to fight, but humans have additional reasons to fight. These reasons include ideals such as freedom, religion, civilizing the conquered, and more. Materialism can also spur wars. World War II lifted the United States out of the Great Depression, because war can have a significant positive economic impact as governments spend money on arms and supplies, and as companies scramble to fill those orders. Companies themselves may instigate or at least lobby for wars. The current conflict in Afghanistan may have retribution for the attacks of September 11th, 2001, as its stated goal, but there some who believe that the actual reason was to secure a route for a future oil pipeline from central Asia to the sea. Whether this is true or not is both open to question and beyond the scope of this book. What is true is that some large corporations and their shareholders involved in producing arms or supplies useful in wartime enjoy massive profits when wars happen.

Crowding by itself is not a reason for conflict. In Chapter 10, I briefly discussed maintenance behavior that is designed to reduce tensions and avoid conflict among group members. Some researchers, including Desmond Morris (*The Human Zoo*), suggest that high population density leads to increased violence, and even outright warfare. This is not the case, at least not in humans under normal circumstances where there are plenty of resources for all (such as a relatively affluent city). It is true that most urbanites and even suburbanites might see more people in one day than our ancestors saw in an entire lifetime because today's populations may be 100,000 times as dense as primordial societies. This may seem like a recipe for disaster, but humans have responded by evolving complex appeasement behaviors whose overall effect has been to reduce the level of violence within a society.

Your chances of being murdered by a fellow member of your society are less today than at any time in history or prehistory, newspaper headlines and "get tough on crime" politicians notwithstanding. You know this if you've ever bumped into someone: Both parties will normally issue immediate apologies, the profuseness of which are directly related to the severity of the collision. People on crowded buses or subways or

Political history is largely an account of mass violence and of the expenditure of vast resources to cope with mythical fears and hopes.
Murray Edelman

walking down busy streets take pains to avoid direct eye contact with others, because staring is a form of challenge. When two people make eye contact, the usual response is a swift greeting, such as mumbling "Hi" or offering a slight nod or other "I'm not challenging you" gesture. This nod is usually downward—a shortening of the body, sometimes with a slight stooping posture thrown in—a clearly submissive gesture. Interestingly, people sometimes respond with an upward nod, which can be interpreted as, "I'm not going to fight you, but I would win if I did." I've already mentioned how Koreans have elevated conflict avoidance to an art form by pressing on after bumping into each other, because an automatic apology is simply assumed. Koreans have some of the highest population densities anywhere, making collisions unavoidable.

This explains why mere population density fails to cause widespread violence. The caveat is that the general population must have at least a reasonable modicum of resources at its disposal. A crowd of people without adequate resources is an entirely different matter. Bloody revolutions and civil wars have started that way.

A Very Brief History of Conquest

Humans evolved in Africa and emigrated to Europe, Asia, North America across the Bering land bridge, eventually reaching South America. The casual observer may wonder why most of the world's successful civilizations (as evidenced by sheer size and/or technological achievements) originated in northern Africa, Europe, and Asia between the latitudes of roughly 25 to 60 degrees north of the Equator. The Egyptian, Babylonian, Persian, Carthaginian, Greek, Roman, Mongolian, Chinese, English, Spanish, Portuguese, Napoleonic, Ottoman, German, and other empires all originated within this area. These empires spurred, and were spurred by, technological achievements so great as to eventually allow the discovery and conquest of the American empires (Aztec, Mayan, and Inca in addition to countless Native American tribes and nations). At no time was any American empire able to cross the Atlantic, much less mount any campaign against Europe or Asia. Why this great imbalance of power?

Geography provides a powerful clue: Eurasia is aligned on an East-West access, thereby creating a temperate climate band

---

The tools of conquest do not necessarily come with bombs and explosions and fallout. There are weapons that are simply thoughts, attitudes, prejudices—to be found in the minds of men. For the record, prejudices can kill and suspicion can destroy, and a thoughtless, frightened search for a scapegoat has a fallout all its own—for the children and the children yet unborn. And the pity of it is that these things cannot be confined to the Twilight Zone.
Rod Serling

stretching for many thousands of miles. This vast temperate zone is home to many types of readily domesticated edible grasses, such as wheat, rye, and barley. Contrast that with the North-South alignment of the Americas: Here, two much smaller temperate bands north and south of the Equator are separated by a mountainous tropical zone with jungles that remain almost impassable to this day. This tropical zone in Central America, particularly the Darien area of modern-day Panama, effectively thwarted any large-scale migration or invasions. If this wasn't enough, *teosinte* (the forerunner of corn) required much time and effort to domesticate. In *The Third Chimpanzee*, author Jared Diamond points out that the average rate of southward migration of people who had crossed over the Bering land bridge was only about one mile per year.

The Americas also lacked what is quite possibly history's trump card: The horse. This one animal has served as a tractor, locomotive, personal vehicle, and troop carrier for many thousands of years. The expression "calling in the cavalry" refers to the tremendous advantages horses brought to warfare. Many infantry brigades have been destroyed (and many others saved) by horse-mounted warriors. By contrast, America had animals such as burros, alpacas, and llamas. These animals are smaller, slower, and less easily domesticated than the horse. Given a choice, it would be probably be better to fight on foot than mounted astride one of these animals.

Widely available, easily domesticated grasses ensured a reasonably steady food supply that agriculture augmented. This steady food supply allowed growing numbers of people to turn their powerful minds away from finding food to other specialties that eventually included metallurgy, which led to better weapons and armor. Horses carried riders and their metallic offensive and defensive implements. Other technological advances included ships and navigational methods that allowed those ships to cross open oceans. In retrospect, the conquest of the Americas by Eurasian people seems almost unavoidable. Had the alignment of Earth's continents been reversed, it would not be difficult to imagine Aztec or Incan conquistadors invading Europe.

The true law of the race is progress and development. Whatever civilization pauses in the march of conquest, it is overthrown by the barbarian.
Georges Anne Besant

Genocide

Was all this bloodshed and deceit—from Columbus to Cortes, Pizarro the Puritans—a necessity for the human race to progress from savagery to civilization?.
Howard Zinn

My dictionary defines *genocide* as the deliberate and systematic extermination of a national, racial, political, or cultural group. Virtually all areas of the globe have seen some form of genocide from Nazi Germany to Rwanda, Sudan, Bangladesh, East Timor, Australia, and the Americas. In my opinion, genocide is distinct from warfare because the latter seeks to attain a specific objective (such as the "liberation" of a country), while the latter merely seeks to kill as many people as possible. So intent was Nazi Germany on exterminating the Jews, that this effort actually diverted resources that could have been used to directly fight World War II.

Some typical patterns of genocide include:

- A militarily superior power invades a weaker nation that chooses to resist. The wholesale slaughters of Native Americans in both North and South America at the hands of invading Europeans and their descendants are perfect examples of this type of genocide.

- A power struggle within a diverse society can trigger genocide, with the targets seen as either blocking the road to power (such as the Tutsis in Rwanda) or as scapegoats (such as the Jews in Nazi Germany).

- Racial and religious persecution is the third major cause of genocide. Examples include the Crusades and quite possibly the Catholic Inquisition.

There is no crueler tyranny than that which is perpetrated under the shield of law and in the name of justice.
Montesquieu

The common denominator in all of these examples is a sense of rightness and entitlement on the part of the perpetrators. For example, cleansing the Americas of dark-skinned, Godless, uncivilized heathens was the right thing to do according to the Europeans, who believed they were bringing the light of material and spiritual progress to these lawless lands. Chimpanzee wars can arguably be called genocides, because chimpanzees have no concept of civilian versus military targets, combatants versus noncombatants, justifiable versus collateral damage, etc. They fight whoever is in their way. Ants discern friend from foe according to genetic differences. The more different the genes, the more likely it is that battle will ensue. Some speculate that the human capacity for committing genocide (literally eliminating different genes) is directly descended from our animal ancestors. Interestingly, genotype difference is a huge factor in human attraction, which we'll examine in Chapters 19 through 23.

The sense of community that inhibits violence against fellow group members provides the foundation for *morality*, which can be loosely said to originate from the sense of fairness derived from the combination of personal emotions and knowing how one's behavior will affect others. Culture helps define which behaviors are deemed good or bad. Few Americans would fail to stop to apologize to someone they bumped into. Few if any Koreans would stop, because they assume an implicit apology. A typical American takes umbrage at an incident that wouldn't faze a Korean. On the other hand, a Korean may be annoyed by the perceived waste of time and effort should someone issue such an apology.

Morality is by its very nature restricted to the "in" group. Outsiders do not normally benefit from a group's morals, except in cases where two or more groups come together to define a universally accepted morality through such means as laws and treaties that specify some form of in-group treatment for outsiders. This has the effect of creating a sort of "über-group," and can have both positive and negative effects. People can travel with ease between many nations (particularly within the European Union) thanks to treaties that allow virtually unhindered passage. On the other hand, many nations have pledged to go to war in the event that one of their allies is attacked. Under normal circumstances, the United States might have no reason to interfere if someone attacks Luxembourg. Thanks to NATO, the United States is obliged to respond to just such an attack as if our home soil had been targeted.

Armed & Dangerous

I've already mentioned that chimpanzees use weapons when attacking other chimps. Chimpanzee wars can be extremely violent affairs; however, they normally involve no more than a few dozen individuals. Furthermore, apes lack the formal training programs and formalized command structures of human armies, which limits the potential damage to a few dead on either side. There is little to no chance that chimpan-

Setting loose on the battlefield weapons that are able to learn may be one of the biggest mistakes mankind has ever made. It could also be one of the last.
Richard Forsyth

zees will become extinct because of their wars, nor is there any chance of any other species dying out for the same reason.

Human combat is a different proposition. Wars have claimed untold millions of lives during the past few thousand years. With some notable exceptions, victory usually went to the army with the most soldiers. The relentless march of technological progress has forever altered that paradigm. From about the 1800s onward, victory has tended to favor the army that could lob the most ordnance at the enemy. Very recent technological developments are now shifting the balance of power to the armies that possess the best information.

The good news is that today's wars are being increasingly fought with precision weapons that hit precise targets to eliminate military threats while minimizing so-called collateral (civilian) deaths and damage. The bad news is that humanity lives in the shadow of the "most ordnance" days, a shadow cast by nuclear, biological, and chemical weapons that are capable of turning the only planet in the entirety of the Universe known to harbor life into yet another lifeless hunk of rock. Tiny armies with zero hope of numerical or informational victory (such as North Korea and terrorist organizations) are actively pursuing these weapons in order to wreak havoc far beyond their current means. The Soviet Union may have lost the Cold War, but the aftermath of victory may be less-than-perfect control over the nuclear arsenal, which may help these tiny armies get their wish.

Weapons are standard fare in human conflict, and few fights to the death occur bare-handed. It is one thing to kill a person with one's bare hands. It is far easier to kill by picking up a stick or a club, even easier to use a knife or a sword. Ranged weapons such as slingshots, bows, and catapults allow killing without the attacker needing to see the target. Firearms can kill with a twitch of a single finger that requires less effort than typing a single word. Modern weapons can be dropped or fired from dozens, hundreds, or even thousands of miles away. Killing has gotten both easier and less personal. Killing someone face-to-face, especially bare-handed, is one thing. Pressing a button that causes people many miles away to die is quite another thing. If you don't believe me, then consider for a moment that hearing about a homicide on your block will

---

A weapon is a device for making your enemy change his mind.
Lois McMaster Bujold

The most potent weapon in the hands of the oppressor is the mind of the oppressed.
Stephen Bantu Biko

probably affect you far more than reading about many casualties in a war being fought halfway around the world.

War has clear evolutionary roots and equally clear evolutionary benefits when fought on the scale we evolved with over millions of years. The potential cost of war has only recently risen beyond any conceivable benefit to our species. For our sake, and for the sake of the untold millions of species that inhabit this one blue planet in the seemingly infinite vastness of the Universe, humanity must find some means to forever lift the threat of total annihilation.

Now the sirens have a still more fatal weapon than their song, namely their silence. Someone might have escaped from their singing; but from their silence, certainly never.
Franz Kafka

Chapter 13

Status at Work

> *Bureaucracy defends the status quo long past the time when the quo has lost its status.*
> Laurence J. Peter

Anyone searching for a distinct entity with its own unique culture, complete with hierarchy, politics, conflict, and pervasive distinctions between positional and personal status and power, need look no further than most companies. The advent of agriculture has freed an ever-increasing percentage of most societies from the daily search for food, and permitted an ever-expanding specialization of skills and professions that led to commercial enterprises. It is also largely responsible for the extreme disparity between haves and have-nots, as we saw in Chapter 11. A hunter-gatherer society relies on all of its members to contribute, with rare exception given to the old and infirm, who can still contribute their knowledge and wisdom. Hierarchies do exist in these societies, and rank does have its privileges; however, the chief's standard of living is not nearly as far removed from those of the rest of the group, as it is in "civilized" societies.

To date, I have worked in a butcher shop, at a stationery store, in the US Coast Guard Reserve, and as a clerk and manager at several retail hardware stores. I started several businesses including a small flight school, and a telecommunications support company, and have been both employed and self-employed as a technical writer and author. As I type this, I am launching a new endeavor to create small business training materials. This fairly diverse background represents the tiniest fraction of all the professions available in today's world. That much is obvious. What isn't so obvious is that all of this seeming complexity hides the fact that life still consists of the same six elements I described in Chapter 2: predator avoidance, group status, sustenance, shelter, reproduction, and death. Our egos may want us to believe that today's cosmopolitan lifestyles are far removed from those of primitive societies, but the illusion is exposed once one looks behind the shell game that is daily life for much of the world's population.

The Profession Deception

Gone are the days of traipsing for hours or days on end in the hopes of finding food for the majority of people alive today. Or are they? Love or hate what we're employed or self-employed to do, the fact remains that the daily hunt for food and shelter is alive and well. Instead of going out to kill a deer and pick some vegetables for supper, I am sitting in a room pressing buttons on a device that is capturing my thoughts into a form that you have presumably already bought (for which I thank you). Your purchase combined with those of many other readers (please tell a friend about this book!) will, with any luck, allow me to purchase food that may or may not include venison. In other words, this book is my way of searching for food and shelter. OK, it's one of several ways, but still...

Getting ahead in a difficult profession requires avid faith in yourself. That is why some people with mediocre talent, but with great inner drive, go much further than people with vastly superior talent.
Sophia Loren

It is very easy to get lost in the world of processing orders, helping customers, designing processes for tracking software revisions, logistics, training, communicating, reporting, and the many details that occupy our minds day in and day out, often for many hours beyond the traditional 40-hour workweek. We busy ourselves with this mountain of information in order to earn resources (profit) for the company we work for (or own), out of which we hope to receive enough resources to cover our own needs of food, shelter, and maintenance (such as health care). Put another way, work has become the modern equivalent of hunting and gathering. We commute to the hunting grounds (offices) every day, hope to bag enough game (profits or other recognized effort), and return to our caves (homes) to rest up for the next day. Whether you're a butcher, a baker, or a candlestick maker matters none; you are hunting and gathering every time you go to work, whether or not you want to admit it to yourself. Everything else, all of the millions of products and services at our disposal today, reflect neither more nor less than the combined efforts of billions of people to feed and house themselves and their families.

The Universal Resource

If all the rich people in the world divided up their money among themselves there wouldn't be enough to go around.
Christina Stead

Our primordial ancestors invested their energy seeking exactly what they needed: plants and animals for food and clothing, and locations and materials for shelter. As I mentioned in Chapter 10, someone blessed with an overabundance of goods might share with fellow group members who would then return the favor. It is also safe to assume that simple barter happened regularly if different people preferred different foods, in the same way that schoolchildren might swap lunches. The increasing specialization allowed by agriculture allowed some people to devote all of their time to growing animals and vegetables while other made clothing, built houses, etc. Barter now became more complex. A farmer might trade some chickens for clothing, for example. Problems arose when one party didn't need what the other had to trade, such as if a chicken farmer wanted to trade with a clothier who already had too many chickens. Both parties were at a loss unless they could find a third party with something else to trade. Growing populations made problems like this overly complex, threatening to bring trading—and civilization itself—to a halt. What to do?

Ever inventive, humankind invented money as a universal means of exchange that freed us from the increasingly intractable problems of barter. The history of money itself includes giant stones, seashells, coins made of precious and semi-precious metals, paper, plastic, and electrons. This last example might seem ludicrous, until you remember that most of your own money exists as nothing more than an entry on a hard drive in one of your bank's many computers.

My dog is worried about the economy because Alpo is up to 99 cents a can. That's almost $7.00 in dog money.
Joe Weinstein

Money soon begat its own complexities. Freed by agriculture from having to hunt or gather their own food, people now had to find other occupations in order to earn money to buy food. I'll discuss this more in Chapters 16 and 17. Money is also the only commodity that one can hoard indefinitely.

Chickens and vegetables don't last forever, and most other commodities either wear out, decay, and/or require increasing amounts of storage (sometimes under carefully controlled conditions). No so with money. Accumulate enough small-denomination units, and you can trade them for one unit of a larger denomination. Imagine being able to trade five leather hides for one of the same type and size worth five times as much. Money is the only commodity than can accommodate this. Thanks to technology, it is also the only commodity that can exist both physically (such as in paper and/or metal form) or virtually (in the form of electronic accounts that are accessible using checks, credit cards, debit cards, and money orders, etc.). This has allowed some people to become unimaginably wealthy, as I described in Chapter 11.

Whenever it is in any way possible, every boy and girl should choose as his life work some occupation which he should like to do anyhow, even if he did not need the money.
William Lyon Phelps

Money has also disconnected people from their work. Before money, success depended almost entirely on each person's skill and wherewithal. She or he who expended the most effort could usually count on the greatest reward. Today, those who expend the most physical effort are often on the lowest rungs of the socioeconomic ladder. These people tend to be the ones who make life safe and comfortable for the rest of us. We get periodic reminders when some of them, tired of their working conditions, stop working and protest for improved conditions. It only takes a few days for a garbage workers' strike to raise a real stink, as well as the risk of an epidemic. When this happens, those in charge usually offer the minimum possible to quell the unrest, and life resumes its normal routine.

I have already argued, and will argue again, that the people who expend the most physical labor should be the ones who receive the best food, shelter, health care, etc. No "primitive" hunter-gatherer society would long tolerate someone who sat around pushing a pencil while demanding the tenderest meat and the ripest fruits.

Tribes within Tribes

At the beginning of this chapter, I noted that most companies are entities unto themselves that qualify as tribes/societies within larger societies (cities, states, nations). That may seem

Society is now one polished horde, formed of two mighty tribes, the Bores and Bored.
Lord Byron

like a pretty far-reaching statement, until one considers the following traits that exist within companies of all sizes:

Companies have cultures that include rules, norms, and taboos. They may also include a certain esprit de corps, values, uniforms, and other cultural phenomena. One lasting idea is the notion that employees and contractors must work in the office, in order to maximize interaction with (and therefore loyalty to) coworkers and the company as a whole. There is very little I do for my clients that requires my physical presence anywhere, yet they routinely request that I visit their offices on a fairly regular basis.

> *Everything that irritates us about others can lead us to an understanding of ourselves.*
> Carl Jung

Business expert and author Michael Gerber speaks of three distinct business personalities: forward-looking *visionaries* who generate new ideas, backward-looking *managers* who try to apply law and order to the goings-on, and *technicians* who live in—and take action for—the moment. Technicians are the ones most likely to quit a job and go into business for themselves. They are also the ones who are most likely not to succeed the way they hoped, because they lack the vision and/or management skills they need to grow a thriving business. In large companies, selection pressures tend to send these three personality types down very different paths. Visionaries may have roles in research and development, and may often found companies. Managers can become executives, and may (depending on many variables) have the shortest route to the top of the socioeconomic ladder. Technicians tend to become self-employed, or remain at the lower end of the corporate and/or socioeconomic ladders.

There are exceptions, of course. Extremely skilled technicians can earn an extremely comfortable living (such as computer programmers during the so-called "Internet boom" of the 1990s). Some of these people may be promoted to management because of their extraordinary technical skills. A few of them can lead effectively, but many of them can't. Ordinary technicians who lack highly sought after skills often have the toughest time of all.

> *When a person can no longer laugh at himself, it is time for others to laugh at him.*
> Thomas Szasz

Managers need not know how the actual job gets done, although they do need at least a solid conceptual understanding. These people need to be excellent leaders, and good with complex problems and logistics. Ask a manager to write programming code or operate heavy equipment, and you may have a disaster on your hands. Assemble enough of even the best technicians for a project sans a good manager, and chaos will soon ensue.

Competition and xenophobia exist at all levels. This can occur between companies, such as rival carmakers, and can even occur within companies with internal departments becoming companies within companies. Some specialization is good. I, for one, am perfectly happy to let my stomach store and begin breaking down food while my lungs oxygenate my blood and my liver filters that blood. I strongly prefer that my intestines have nothing to do with locomotion (and I mean that both literally and figuratively). You get the idea. The difference between internal organs and corporate departments is that internal organs (departments) cooperate to benefit the entire body (company), while departments in corporations can and do act against each other. Spend any time examining most any company, and you'll probably walk away thinking it a miracle that the lights come on in the morning—a thought that has nothing to do with the physics of the matter.

When you make a world tolerable for yourself, you make a world tolerable for others.
Anais Nin

Society may dislike cheaters as a general rule; however, I've seen more than one company where people routinely take credit for other people's work, wrongfully accuse others, and otherwise flagrantly place their own interests above others. The extent to which the company tolerates and even rewards this behavior speaks volumes about the company's values (or at least about its fear of a wrongful termination lawsuit).

Reciprocity tends to be watched like a hawk. Who did favors for who, who didn't return a favor, which department helped which other department, which department needs to pay for a cross-functional expense—the list goes on and on. The rules for engaging in and tracking reciprocity can be complex in the extreme. More than once I've helped someone in a different department, only to be warned against doing so for various accounting and/or political reasons.

You must give some time to your fellow men. Even if it's a little thing, do something for others—something for which you get no pay but the privilege of doing it.
Albert Schweitzer

Status cues abound at companies, from clothing to offices-versus-cubicles, job titles, pay scales, bonuses, and more. Companies place heavy emphasis on their hierarchies, and go to great lengths to show them off to anyone (internal or external) who cares to have a look. I've already talked about the extreme disparities between those at the top of the ladder and those at the bottom. Only in an agricultural society that uses a uniform means of exchange (money) can people earning over 1,000 times as much as an average worker exist. Is it any wonder that agriculture spread so slowly, and that some societies still refuse to embrace it?

There are two kinds of people, those who finish what they start and so on.
Robert Byrne

Jockeying for position is another key feature of many companies. It never ceases to amaze me what people will do to stand out in hopes of receiving a promotion, bonus, or other praise or recognition. Some of the hours people put in at the office make me tired to even think about. Behavior like this is not only allowed, but encouraged, thanks to such mechanisms as overtime laws that don't apply to certain types of workers. It is not uncommon for someone putting in these long hours to actually earn less than someone with a lower base salary who qualifies for overtime.

The concept of "hard work" is also used to motivate people to spend that much time at work, in the vein of "So-and-so earned this promotion thanks to her hard work." The never-overtly-stated-but-never-subtle message is that people who work to live will never achieve as much as people who live to work. The fatal flaw with this message and its underlying expectation is the definition of work as effort. Companies tend to view the people who "work hardest" as those who work the longest. Physics, however, defines *work* as results. Someone who takes four hours to do the same job as someone who takes twelve hours has done the same amount of work. Chapter 5 of *The Enlightened Savage* explains this concept in much more detail.

The secret of a good life is to have the right loyalties and hold them in the right scale of values.
Norma Thomas

Loyalty is a potentially disturbing concept that both societies and companies employ. A worker (citizen, subject) is expected to be loyal to the company, state, etc. Meanwhile, the company rarely shows that loyalty back to the workers who built and maintain it. On the contrary, companies will go to great lengths to avoid having to keep their explicit or implicit promises to their workers. Layoffs and downsizing occur too regularly to notice, except when large companies shed an unusually large number of workers. This often occurs with zero notice. I know. I've been laid off more than once. People can be laid off, fired, or offered the "opportunity" to keep their jobs by moving to remote locations in order to avoid paying pension or other retirement benefits. There are too many other examples of one-sided loyalty for me to list, but you get the general idea. This is a fairly recent phenomenon. As late as the 1950s or even 1960s, a man (usually) could reasonably count on staying with the same company for his entire career and enjoying a comfortable retirement. Fast forward to the new millennium where a now former CEO publicly said that people have no inherent right to a job!

I am not saying that money and commerce are evil, nor am I advocating any radical departure from our generally capitalistic methods, for they alone of all the systems humankind has experimented with hold the greatest promise of true success and freedom for the greatest number of people. Also, there are a great many companies that truly do honor their workers and go out of their way to return the loyalty. I am saying that the advent of agriculture and money, along with the related concept of individual ownership, have paved the way for egregious exploitation on a hitherto unprecedented scale.

> *Bureaucracy defends the status quo long past the time when the quo has lost its status.*
> Laurence J. Peter

Women at Work

Sexual division of labor is both common and necessary in many species from fish to mammals, and certainly humans. Among apes, the male does the hunting and protecting, while the female raises the children. Humans have long had a similar division of labor: The man hunts and defends the tribe, while women forage, raise children, and tend the home. I liken this system to the American effort during World War II: We could not possibly have won the war without soldiers in the field doing the fighting. We also could not have won without the "home front" that built and shipped the vast amounts of supplies needed by those soldiers. On a more personal level, why hunt and defend the home if there is no home to defend? Why must anyone do all the work? Sexual division of labor makes clear evolutionary sense for many animal species, and for humans as well.

Sexual division of labor makes so much sense, in fact, that humans are physically built for their roles. Men tend to be more athletically capable (good for long hunting forays) and can take a fair amount of punishment to the chest (the single largest exposed mass on the body) without much damage. On

> *Whatever women do they must do twice as well as men to be thought half as good. Luckily this is not difficult.*
> Charlotte Whitton

> *If you have any doubts that we live in a society controlled by men, try reading down the index of contributors to a volume of quotations, looking for women's names.*
> Elaine Gill

the contrary, hitting a woman in the chest is the rough equivalent of hitting a man in the groin. A woman's wide hips are less well adapted for running. Male hormones, testosterone in particular, invite aggressive behavior. Estrogen encourages more nurturing and less aggression. Men tend to think in black and white, while women are much better at thinking in shades of gray. Women have better color perception and fine motor coordination, the better for seeing and picking ripe fruit high in a tree with an infant clinging to them. All humans share a common genetic legacy, and many of the same bodily structures (organs, muscles, etc.). That sameness (what a car maker would call a "common platform") ends when it comes to the adaptations that favor certain functions over others. In these areas, there is nothing equal about the way men and women are wired. Biologically, we are very different creatures.

That said, I don't for a moment believe that men are superior to women in any way, or vice-versa. Each gender has its inherent strengths and limitations that together have helped humankind flourish. I believe that my gender has made the fundamental error of devaluing the critical role of women in society, to the point of believing that so-called "original sin" came from women, whatever that monstrous concept is. Men have done such a good job at this, that women finally decided to compete with us for equal treatment on our terms. Women's suffrage, feminism, and the rush of women into the workforce ensued, with the result that women are now dying from the same stress-related disorders as men, and starting to erode their statistically longer life expectancies.

> *The thing women have got to learn is that nobody gives you power. You just take it.*
> Roseann Barr

The rush of women into the workforce has also had some interesting economic effects. Introducing more people into the economy to compete over the same fixed quantity of goods drives up the cost of those goods. I have heard some speculation that the large influx of women to the workforce helped create a sort of self-fulfilling prophesy, in that many households today are struggling with two incomes to maintain the lifestyle that earlier generations enjoyed on a single income. I don't know how much of this is due to simple inflation. I do know from my own experience that earning two incomes brings additional expenses in the form of higher taxes and outsourcing domestic functions like housekeeping, gardening, cooking (more and more meals are packaged ready to "heat and eat"), and even child care. Two people earning $100,000 each will not enjoy a $200,000 lifestyle.

I am not saying that a woman's place is barefoot and pregnant in the kitchen. Today's technology allows all kinds of arrangements from reverse households (where the man stays home and the woman works) to two part-time workers and, yes, a single income. My comments are not in any way intended as a slight against women for wanting to escape their traditional role. I am saying that all work has value, be it in an office or vacuuming the living room rug, in the boardroom or the nursery. I am saying that we might do well to choose to value those who tend home and hearth—whichever gender they may be—just as much as we value our corporate warriors. Restoring the sexual division of labor while taking pains to give equal credit to both sides of this division may go a long way to making life a lot simpler and a lot more enjoyable.

In my never-humble opinion, of course.

God gave women intuition and femininity. Used properly, the combination easily jumbles the brain of any man I've ever met.
Farrah Fawcett

Chapter 14

Status at Home

Having discussed status and hierarchies, both in general and at work, it is only fitting that I conclude Part III of this book with some thoughts on how status plays into home life. Anyone who has ever lived with one or more people can attest that domestic life offers no respite from status and hierarchy, nor is it necessarily a refuge from the many conflicts that occur each day. On the contrary, some conflict in the home is both predictable and unavoidable, for reasons that include sexual division of labor and the uneven investments parents making in raising children, not to mention genetic factors. Thank goodness for friends!

Sexual Division

Women need a reason to have sex. Men just need a place.
Billy Crystal

I firmly believe that "women's work" is a very real thing driven by biological imperatives that are hard-wired into our genetic code. Yes, this is a sexist remark, in that I am *discriminating* (making fine distinctions between) based on gender. This should not be confused with *chauvinism* (claiming the relative superiority of either gender). To me, the concept of gender-based roles falls squarely in the "separate but equal" camp. By this, I mean that men and women are built for entirely separate tasks, both of which are equally necessary for maximizing the odds of successful reproduction.

Why Women?

Sexual division of labor exists because our children are born completely helpless and take a long, resource-intensive time to mature, as you'll recall from Chapter 8. It is simply impossible for a human mother to leave her infant behind while she goes off to hunt. Unable to feed itself, the infant must unmistakably communicate its needs to the mother. Nothing says "easy lunch" to a predator quite like a crying infant. Older children can walk on their own and gather their own food with some help, but remain vulnerable until reaching sexual maturity. Someone must therefore remain with the child at all times.

I've done my share of hiking with Logan, starting when he was too small to walk. His weight on my shoulders slowed me down, particularly when he would fall asleep leaning to one side in his carrier. His stamina has increased many fold since then, and he can usually take on steep 5-mile treks. I say "usually," because it's getting harder and harder to carry him along with my normal burden of food, water, and warm clothing, when his strength gives out partway through. It's like driving with the brakes on, this from someone who isn't consciously worried about predators, foraging, or finding shelter. Mothers living before the days of waterproof-yet-breathable synthetic fabrics, ergonomic child-carrying backpacks, packaged rations, and forests cleared of most carnivorous animals had truly tough going.

Why mothers? The wide hips that allow their babies' large skulls to pass through during birth come at a price of reduced speed, while men don't have that impediment. A woman's sensitive breasts resting squarely in the middle of her largest bodily mass are easy—and extremely painful—targets, making women less suited for fighting than men. It makes little evolutionary sense to have one parent give birth and the other nurse the infant, because this would increase the vulnerability of both genders, with no discernible benefit. As long as one gender is slower and more vulnerable, having that gender remain at home to care for the young while the other hunts makes the most sense. The same adaptations that make childbirth and feeding the child possible also make ongoing survival more difficult. In humans, women fall into this category.

Male and female represent the two sides of the great radical dualism. But in fact they are perpetually passing into one another. Fluid hardens to solid, solid rushes to fluid. There is no wholly masculine man, no purely feminine woman
Margaret Fuller

How close the sexes sometimes come to one another. It is as much a matter of behavior and the sphere in which they move that separates the masculine part of humanity from the feminine.
Elizabeth Aston

Chapter 14
Status at Home

> *Grown-ups never understand anything for themselves, and it is tiresome for children to be always and forever explaining things to them.*
> Antoine de Saint-Exupery

Win-Win

Children require an astonishing amount of food. At only ten years old, Logan goes through groceries with gusto. Feeding him is a bit like ceaselessly shoveling coal into a ship's boiler. I can only imagine the carnage that will ensue once he hits puberty! My point is that a lone human in the wilderness must devote significant time and energy to finding food, as any survival show will demonstrate. Finding enough food for two while burdened with a heavy, hungry, and noisy child that both scares prey and attracts predators is a tall order indeed. It stands to reason that having one parent dedicated to finding food is a great idea, especially when that food comes in the form of densely packed calories rich in protein and electrolytes. If the hunter has a clean, dry, and warm place to come home to, then so much the better. The hunter benefits from having a home worth hunting and fighting for where he can return to rest and recharge, and mother and children benefit from the food in addition to their own foraging. It truly is a win-win scenario.

The proverbial bacon Dad brings home has another benefit: More food allows earlier weaning, which allows the mother to have more children in a given time frame than a woman lacking this support. Having more children boosts the odds of passing on genetic material to another generation, and thereby winning the great game of life. This kind of collective, mutually dependent living allows both partners to do more—and with greater efficiency—than they could alone. It also requires both partners to have a stake in the outcome. A woman must be able to rely on a man to share the fruits of his labors, and a man needs to be at least reasonably certain that the children he is feeding really are his own.

Monogamy

> *Bigamy is having one wife too many. Monogamy is the same.*
> Oscar Wilde

Welcome to the wonderful world of *monogamy*, where a woman makes herself available to one man and one man only in order to ensure that the man will have a vested interest in both her well-being and any children. It is no coincidence that human males have smaller testes than our freewheeling chimpanzee and bonobo cousins, who have zero concept of monogamy. For them, the biggest balls win, because they can out-produce the competition and increase the odds of winning the evolutionary jackpot by securing a genetic legacy. Monogamous humans with slow-growing children have no need of large testes because our sperm need not compete with that of other males.

This may sound good in theory, but life isn't quite that simple.

Uneven Investments

Man inseminates woman. From a purely biological point of view, his job is done. Should something happen to the mother and/or the unborn child, the man is free to inseminate another woman. The woman, by contrast, must devote an ever-increasing share of her own resources toward raising what is essentially a parasite in the form of a child that relies on its mother to survive. The more the mother provides, the more the child needs. Imagine loaning money to someone who needs ever-increasing infusions of cash in order to possibly repay the growing mountain of debt. Should you keep on loaning money on the possibility of recouping your investment or cut your losses? That is the question every mother asks herself, whether she's aware of it or not. The high percentage of babies that end up miscarried, aborted, adopted, abandoned, or killed bears witness to this questioning. The mere act of bringing a baby to term inside the womb carries potentially fatal consequences for the mother, should complications arise.

Goodness is the only investment that never fails.
French Proverb

Meanwhile, as I've already mentioned, the man is free to inseminate other women. There is no hard biological need for him to stick around to help the mother raise the child, especially in modern times where babysitting and groceries are more available than they've ever been at any time in our evolutionary history. A child can theoretically survive without any paternal involvement whatsoever, and many children do just that. The father's incentive to help comes from both his confidence that the children in question really are his, and the increased childbearing potential the mother gains through his support. You may think this a flimsy foundation on which to build a two-parent life, but the male instinct to stay with a single woman is strong enough to affect even those relationships where children aren't part of the picture. My partner Jennifer

If there were no bad speculations there could be no good investments; if there were no wild ventures there would be no brilliantly successful enterprises.
F. W. Hirst

and I neither have nor want children with each other. Our bond is as monogamous as they come.

If a woman's chief investment in the child-rearing equation is physical and the man's economic, then logic demands that men should be attracted to women with certain physical attributes, while women should be more discerning about the male's ability to provide for her and her children. This is precisely what happens, as we'll see in Chapter 21.

Investment vs. Power

> *Be bold and mighty powers will come to your aid.*
> Basil King

Men and women have very uneven reproductive investments that benefit from a sexually-based division of labor, as we've just seen. Women tend to be physically smaller, weaker, and slower than men on average. This is true for humans, chimpanzees, and bonobos. Still, there is nothing intrinsically stopping women from being in charge. Bonobo societies are matriarchal, while chimpanzees and the vast majority of humans are patriarchal. Bonobos enjoy a very peaceful lifestyle, where conflicts are usually resolved quickly with no harm done. Chimpanzees endure a more violent lifestyle, where conflicts can lead to serious injury and even death.

Most human societies tend to follow the chimpanzee's lead. Even so, there is no intrinsic connection between women's relative powerlessness and their preference for economically viable mates. On the contrary, my guess is that a woman's power and drive to find an economically viable mate may be inversely proportional to some extent. A woman who is independently wealthy or has a solid career may be able to overlook a man's financial shortcomings. A woman who lacks the means to raise a child on her own may be more driven to find a good provider. One interesting exception can occur among very poor women. I'll discuss this more in Chapter 20.

Fundamental Errors?

> *An expert is a person who has made all the mistakes that can be made in a very narrow field.*
> Niels Bohr

I stand behind what I said in Chapter 13: I believe that men have made the fundamental mistake of devaluing women, their contributions to society (and the species), and the tremendous amount of effort required to sustain any household. Only in 2001 was the United States Labor Department studying how to include housework as part of measuring the nation's Gross Domestic Product (GDP). The Genuine Progress Index for Atlantic Canada (report prepared by Ronald Colman, Ph. D) estimates that replacing housework with pay would add $275 billion dollars to the Canadian economy. According to this report, non-employed single mothers work 50 hours per week, while working mothers routinely work 73 hours per week. Salary.com placed a "salary equivalent" value of $134,121 on housework for stay-at-home mothers in 2006. The actual numbers are debatable. What is beyond debate is that housework, raising children, and other "women's work" has tremendous value. How men could devalue and demean this sort of contribution is both baffling and inexcusable in this man's never-humble opinion.

Women, on the other hand, fought long and hard for the equal treatment they deserve, finally winning the right to vote, and being accepted into the workforce in droves. Whole advertising campaigns launched telling women that they could have it all: A solid career, children, and a rewarding home life. The actual results have been less than spectacular. Women are increasingly dying of the same diseases as men, because of the added stresses of having jobs. Children are sent off to day-care centers, sometimes for eight or more hours per day. Despite this, women are still expected to perform housework on top of their other duties. Even worse, many families have reached the point where they can no longer maintain their lifestyle on a single income. Or can they?

I once added up the potential costs of full-time child care, housekeeping, second car, dining out, etc. and arrived at a figure that's just over half my former wife's current gross salary. 30% of that salary went to taxes, leaving about 20% in net income gains per month. I can't imagine anyone who would voluntarily work their current job for 20% of her or his current salary! My calculations don't factor in other potential costs such, as the potential inflation caused by massive influxes of workers to the economy and increased demand for the same overall supply of goods and services. I also didn't factor in potential savings from economizing elsewhere in the monthly budget. If you currently rely on two incomes, then you may want to perform the same calculations for your own situation. The results may startle you.

To avoid situations in which you might make mistakes may be the biggest mistake of all.
Peter McWilliams

Then again, maybe they won't. Every situation is unique. For example, my former wife and I are fortunate in that in our son currently attends school during the same hours that she is at her teaching job. There is no extra car payment, and I can spend lots of time with Logan thanks to my self-employed status. We therefore realize a significantly higher return from her work, thanks to our situation. I am clearly not saying that two-income situations are wrong. I am saying that you need to look beyond the extra salary at the added costs in both dollars and quality time, then make an informed decision. You probably have a lot more flexibility than you think.

I am also not saying that a woman's place is at home. A "reverse nuclear family" where the husband stays home and the wife works can be a perfectly viable solution, as can a dual-income situation. Still, the sexual division of labor exists for solid evolutionary reasons. Men should not devalue or oppress women for their biologically designed roles, and women should not feel that their only path to equality lies in emulating men. Sure, women may make better fighter pilots than men. What exactly does that prove? And why do women feel the need to prove that?

Conflict at Home

The male is a domestic animal which, if treated with firmness, can be trained to do most things.
Jilly Cooper

Home sweet home is a wonderful idea, but the truth isn't nearly that tranquil. Here are just a few examples.

In the Womb

In Chapter 8, I explained how conflict begins before birth with the fetus looking out for its own interests, while the mother is evaluating the odds of the fetus being able to pass along her genetic legacy. Medical technology has enabled women to consciously make the same evaluations their bodies have been making automatically for millions of years, and opt to terminate the pregnancy through an abortion procedure. The only difference between spontaneous abortion (miscarriage) and deliberate abortion is the conscious awareness and choice behind the action. Either way, a failed pregnancy is the natural result of a decision process that finds the fetus lacking.

Weaning and Walking

The next major conflict occurs around weaning, as the child is increasingly expected to fend for itself, instead of simply nursing. Around this time, the child's increasing mobility and natural curiosity foster a growing need for independence, while at the same time needing the reassurance of parental presence and involvement.

Sibling Rivalry

Siblings are another source of conflict. Every sibling carries some percentage of each other's genetic material, but only an identical twin is 100% genetically identical to the other. Here again, the competing needs of each individual to pass along her or his genes comes into play as each sibling vies for food and status within the family. Helping parents raise siblings is a good way to help ensure that at least some of one's genes get passed on, but that fractional benefit can come at the expense of delaying the ability to reproduce on one's own. Siblings also tax parental resources, such the amount of time and attention that can be lavished on any one child. Firstborn children are often accorded special privileges and attention. The youngest sibling often receives extra attention, such as diapering and feeding, while older children take care of themselves. Middle children are the ones who usually receive the least attention, because they lack both the firstborn's status and the last-born's needs. This can result in a middle child making extra efforts to gain attention. I've witnessed this firsthand in several families, including some very close friends of mine. One of these friends once expounded at length about an argument he was having with his older (firstborn) brother and his frustration at feeling unheard and misunderstood. He wanted his older brother to sit down and listen to his side in exquisite detail. I remember advising my friend repeatedly to just let it go, only to realize much later that he couldn't, for all of the reasons I've just described.

Such seems to be the disposition of man, that whatever makes a distinction produces rivalry.
Samula Johnson

Parents vs. Children

Conflict between parents and children is also part of the scheme of things. Think about this: Every resource that goes into feeding and raising a child is one less resource that the

The reason grandparents and grandchildren get along so well is that they have a common enemy.
Sam Levenson

parents have for themselves. This is a fine line to walk. Allocate too many resources to the children, and the parents' own fitness will be reduced, potentially impacting their ability to care for their young and thus pass on their genes. Too few resources, and the children will suffer, with consequences for the parents. It really is a fine line. It is even possible that some child and parent attributes might have evolved to compete with each other.

> *We are the people our parents warned us about.*
> Jimmy Buffett

The analogy of loaning increasing resources to someone with the potential for payback versus cutting one's losses applies here as well. The lifeboat adage "women and children first" makes evolutionary sense, because one man can impregnate many women, while each woman can only experience one pregnancy at a time. Sacrificing men to save women therefore makes sense, because it has the least long-term impact on the population. One could argue that making an adult consumes far more resources than making a child, and thus the children should be sacrificed. This argument ignores the fact that children carry their parents' genes, which means that the parents have already won the great game of life. From an evolutionary perspective, saving the women and children is the best thing to do. Thank goodness that modern ships are required to carry enough lifeboats for all!

Spousal Conflict

> *There's only one way to have a happy marriage and as soon as I learn what it is I'll get married again.*
> Clint Eastwood

Few comedic routines are complete without at least a few lines devoted to marital strife. Untold thousands of articles discuss the most common topics of conflict between spouses, and advice columns routinely offer advice to people who are at odds with their significant others. These address the stated cause of the problem (such as money), and some even delve into the feelings and motives behind the scenes. Few sources designed for mass consumption approach marital conflict from an evolutionary perspective.

Couples with children share the goal of successfully raising their young and passing along their combined genetic material. There's just one problem: The mates themselves share no genetic material under normal circumstances, and a child carries only 50% of either parent's genes at best. Each mate's investment in the couple's children confers a fitness benefit on the other mate. This paves the way for one or both mates to take undue advantage of the other. The male's natural instinct is to seek out as many mates as possible to pass on as much of his genes as possible. The woman's incentive is to throw all she has into raising her current child, because she can't hope to compete with the male in sheer numbers, and must therefore focus on every child. It's the old "quantity versus quality" conundrum. Even the most monogamous male has thought about investing some resources in an extramarital encounter... and any male reading this who never had before, just did.

It doesn't end there. Remember when I said that animals (humans included) are far more likely to help their kin than non-relatives in Chapter 4? Under normal circumstances, each partner in a couple has their own family of blood relatives. Devoting all of one's attention to one's own child helps a non-relation (the other parent) pass along her or his genes, while possibly limiting the ability of other family members to pass along at least some of one's own genes. Strife over the amount of time and energy spent on family is a common refrain in many, many marriages. If you've ever argued about your in-laws, or seen or otherwise know about such an argument, then you know exactly what I'm talking about.

I recently purchased a new laptop computer. I did my homework, researching brands, prices, and value before placing my order. Nevertheless, I found myself eyeing a competing model longingly within moments of clicking the "Place Order" button on the seller's web site, because it suddenly occurred to me that I may have missed out on an even better value. This *buyer's remorse* isn't limited to consumer electronics. One or both parties in a marriage or other pairing may become convinced that they can get a better value for what she or he has to offer by leaving the relationship. Cases where both parties want out have the best chance for peaceful resolution. My former wife and I both decided to end our marriage, and we're better friends today than we've been in many years—if not ever. Cases where one party wants out are where problems arise. A man I know decided to leave his marriage. The cost in time, energy, and money spent on the divorce proceedings themselves could literally have purchased a comfortable house, and that's not including the division of marital property. Women who decide to end a relationship risk violence or other pressure not to go through with the breakup.

There is no such thing as "fun for the whole family."
Jerry Seinfeld

If you ever start feeling like you have the goofiest, craziest, most dysfunctional family in the world, all you have to do is go to a state fair. Because five minutes at the fair, you'll be going, 'you know, we're alright. We are dang near royalty.'
Jeff Foxworthy

Many men can make a fortune but very few can build a family.
J. S. Bryan

Power is another huge source of conflict. The woman invests more in child-rearing, but most societies recognize the man as the head of the household. Women even give up their family names when married, and the typical Western family wedding ceremony has the woman's father "giving away" the bride in a procedure not unlike handing off a piece of property.

No discussion about spousal conflict would be complete without at least mentioning sexual infidelity. Chapters 19-23 cover sexuality and sexual issues in depth.

As if this wasn't enough, one or more parents may also have children from a previous partner. Children born from the current relationship reflect a mutual investment, and more than one couple has remained together "for the kid's sake." Children born from past relationships only carry one parent's genes, and the other parent has no evolutionary incentive to invest anything in them. In fact, the step parent has every evolutionary reason to prevent resources from going to the stepchild, for the same reason that male lions kill cubs they haven't sired: to increase their odds of passing on their own genes. Tales about children doomed to live with rotten stepparents abound all over the world, and are based in real-world truth. Several studies have shown that stepchildren are far more likely to be abused than biological children regardless of external factors such as socioeconomic status. Stepchildren are a large source of conflict.

I must take care to distinguish stepchildren and adopted children, where both parents opt to rear a child that is not related to either of them. Overall, adopted children fare very well. Adoptive parents allegedly commit child abuse in 1% of reported cases while comprising 2 to 4 percent of the general population, according to the United States Census Bureau. According to the United States Department of Health and Human Services, 12 children out of every 1,000 aged 18 or younger were abused in 2005. The figure for adopted children is therefore far lower out of the same 100,000-person sample.

God gives us relatives; thank God we can choose our friends.
Ethel Mumford

Most people have a strong aversion to violence against genetic relatives. That aversion does not extend to spouses, who normally aren't genetically related to each other. Approximately 7% of married people experience spousal abuse (Statistics Canada, 2004).

Friends

Imagine someone who (probably) isn't related to you, who (probably) shares little if any of your particular daily stresses, but who cares about you and has your back. I am blessed with a large network of close friends, and have no idea what I'd do without them. People who depend on each other tend to cooperate better than people who don't have any such bonds. Friends are important to animals as well. Apes spend about 10% of their days grooming each other, and otherwise building and maintaining social bonds. Females with the best social connections successfully raise more offspring. Primatologist Frans de Waal was once greeted by a happy bonobo, who recognized him after 20 years of separation. Giraffes graze near friends about 15% of the time.

My informal observations indicate that most friends are of the same gender. One female friend of mine remarked some years ago that it's impossible for men and women to be in each other's presence under any circumstances without everyone involved at least momentarily contemplating the prospect of having sex with everyone of the opposite gender (and a few of the same gender since not everyone is heterosexual). This may partially account for the popularity of gender-segregated clubs and other social networks. As for me, I've been alternately fascinated by—and wishing I'd never heard—that comment.

From letting one vent about their rotten day to helping in all manner of activities from moving to gaining status and power, the Beatles were right: We really do get by with a little help from our friends. *The Social Savage* (coming 2014) explores friendships in more detail.

It takes a great deal of courage to stand up to your enemies, but even more to stand up to your friends.
J. K. Rowling

PART FOUR

Three Hots and a Cot

182 | *The Natural Savage*
Discovering the Human Animal

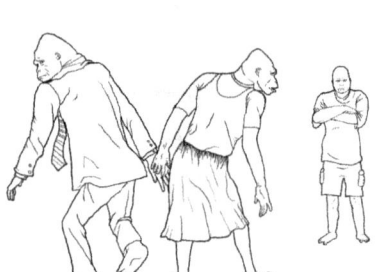

Chapter 15

The Omnivore's Dilemma

> *Health food makes me sick.*
> Calvin Trillin

Shifting climates forced our ancestors to leave the dwindling primordial forests and seek their fortunes on the vast African grasslands. Fruits, leaves, and bark are in short supply on the savannah—as is most humanly edible vegetation—because we simply cannot digest most of the grasses and bushes that grow there. What the savanna does have is vast herds of grazing animals for the taking. It also has plenty of predators to eat those grazers that can also outrun and outfight humans. Partaking of the feast required our ancestors (who had evolved to eat fruit in the trees) to compete with predators that had evolved to eat meat on the grasslands.

Humans already had some experience eating meat. Chimpanzees eat other chimpanzees, and also occasionally hunt small animals. It is likely that early humans share the same legacy. This innate ability to eat such a varied diet is what saved the human species. Exercising that ability required us to either out-graze animals evolved for grazing or out-kill animals evolved for killing. The combination of strong hierarchical social bonds forged in the forests and artificial weapons closed the gap, followed much later by agriculture. Humans survived and flourished. Some speculate that encounters with wolves—who we eventually domesticated into dogs—had a profound impact on our own evolution because one can draw many parallels between human and wolf societies.

Complex Larders

An animal whose diet consists of a single thing faces an absurdly simple decision about what to eat: If the object in question looks, feels, smells, and tastes like lunch, then it can probably be eaten with impunity. This simplicity carries a price because any disruption in habitat can prove disastrous. An animal capable of eating many different things enjoys at least some insulation against habitat disruption, but this carries the risk of inadvertent poisoning. The balance still favors the latter animal, because poisoning tends to be an individual risk, while losing one's only food source can threaten the entire species.

If it weren't for Philo T. Farnsworth, inventor of television, we'd still be eating frozen radio dinners.
Johnny Carson

This predictable outcome is precisely what is happening today: Climate change caused or exacerbated by human activity, combined with large-scale habitat changes thanks to logging, mining, agriculture, and urbanization, are pushing highly specialized species towards extinction. Forest-dwelling apes are vanishing, just at the time when science is beginning to understand them and, through them, understand our own species. Meanwhile, generalized species like raccoons and rats are thriving. In general, the more things you can eat, the better. A generalized diet may also mean a wider possible range of habitats. For example, rats are found throughout the world.

But what about the poison? Michael Pollan, author of the book from which I took this chapter's title, makes the convincing argument that an evolutionary defense against poison would require complex mental abilities, such as memory, keen observation, rules of thumb about what's good and bad to eat, and how to experiment with unfamiliar foods. I'll get back to this in a few moments.

Food supplies can also drive other behaviors. Animals that rely on a single food source may benefit from being solitary, because any local disruption will affect fewer animals. On the other hand, animals whose food comes in abundant concentrated sources (such as fruit-laden trees) may benefit from being social, for the many reasons discussed throughout this book. This is also true for animals whose food is literally too big for any one of them to down, such as wolves chasing elk.

I don't know a better preparation for life than a love of poetry and a good digestion.
Zona Gale

Finally, what we eat can drive how we perceive the world around us. Ripe sweet fruit tends to adopt a highly contrasting

Chapter 15
The Omnivore's Dilemma

Vegetables are interesting but lack a sense of purpose when unaccompanied by a good cut of meat.
Fran Lebowitz

color, as if to advertise its presence by practically screaming "Eat me!." This is exactly what the fruit is doing, because the fruit's species benefits when animals eat the nutritious fruit, and scatter seeds and fertilizer around the forest in their droppings. It should go without saying that fruit tends to be rather stationary, and that nighttime darkness makes it hard to see. One would expect an animal that depends on fruit to have excellent color vision and to be active during the day. Nighttime vision would be substandard compared to other animals. The ability to detect the subtle colors and patterns of something that didn't want to be found (such as a prey animal) would also be compromised.

This is exactly what happened with humans. We have excellent daytime color vision, and poor night vision. Our superb color perception makes it hard to spot hidden animals. I've already mentioned experiments where color-blind soldiers can readily spot camouflaged enemies, because their inability to perceive color forces them to compensate by developing other abilities. Predator animals tend to be at least somewhat color blind. Small motions also escape human detection. Contrast this with hawks that can spot a rodent moving beneath ground cover from hundreds of feet up. Humans do, however, have 3D vision that provides excellent depth perception, thanks to eyes that are arranged in predator-like fashion—just the ticket for swinging through trees in search of fruit!

The old adage about being what we eat is true on many more levels than normally imagined. Modern humans are a bit of an exception to this rule, because our diets and habitats extend far beyond what our external anatomy might suggest.

Picky Eaters

It's so beautifully arranged on the plate—you know someone's fingers have been all over it.
Julia Child

The mere ability to eat danged near anything that moves (and a wide assortment of things that don't) would be almost worse than useless if it didn't come with some rules about how and when to exercise that ability. On other words, just because you may be able to eat something doesn't mean that you should. As paradoxical as it may seem, being picky about the foods one eats is an essential part of being an omnivore. Remember that the ability to eat a wide variety of potential foods raises the risk of accidental poisoning.

In theory, one would expect an animal in this position to strongly prefer foods that it has eaten before without incident. Next on the list are those foods that others of the same species are known to eat without incident. This animal will also be open to experimenting with unknowns in dire circumstances. And finally, there are some things it won't eat under any circumstances whatsoever. For example, no human in her or his right mind would eat feces or anything reminiscent of feces. This hierarchy of preferences minimizes the risk that eating something will prove harmful or fatal. In other words, this is nature's built-in poison control system.

Vive la Cuisine

Even the most robust poison control system of preferences and rules leaves the culinary door wide open for the enterprising omnivore. Humans encountered new foods in every corner of the world, which gave rise to the many local and regional cuisines available today. We simply could not have spread across the globe without being able to enjoy such a wide variety of foods. Our range of potential edibles is so vast, that it is entirely possible for people of one culture to enjoy foods that utterly repulse people of other cultures. I'll never forget my apartment in San Jose, California, a huge cultural melting pot. More than once, the smells of neighbors' cooking had me literally retching and gasping for air as I hurriedly traversed the shared corridor between units. I can only speculate that my own cooking, which had me salivating in happy anticipation, returned the favor. Let us not forget that the people whose cooking odors disgusted me were presumably just as pleased with their efforts as I (usually) am with mine.

My theory is that all of Scottish cuisine is based on a dare.
Mike Myers

Which brings me to the durian fruit. This large, spiny, thick-hulled fruit has a stench that travel and food writer Richard Sterling describes as "pig shit, turpentine, and onions, garnished with a gym sock." I've smelled durian, and Mr. Sterling's description is positively charitable. The smell is literally that of rotting flesh, which attracts carrion-eating species to help spread the fruit's seeds. Humans are biologically wired to reject such smells, because they convey an extremely high risk of poisoning. So powerful is the stench of durian, that many cities ban the fruit from crowded public places, such as subways and buses.

Whatever prompted the first human to pop a hunk of durian in her or his mouth over 10,000 years ago is utterly beyond me. Still, the person strong enough to overcome natural revulsion to actually try durian is rewarded with a flavor described by British naturalist Alfred Russel Wallace as "rich custard highly flavored with almonds." I've tried durian. Mr. Russel's description is spot on. This is a clear example of someone (me) eating something that I know others eat without incident. If I didn't already know that people eat durian, I would probably starve to death if stranded in an area overflowing with the stuff... and no, I am not exaggerating in the slightest. Disgust is nature's way of preventing poisoning.

Confronting the Unknown

I aimed at the public's heart, and by accident I hit it in the stomach.
Upton Sinclair

Imagine for just a moment that you are stranded in some remote area with little hope of immediate rescue. You must eat if you are going to survive. You know that you are capable of digesting a great many things, but have no idea which of the many things around you might be good—or bad—to eat. What do you do?

Your first step might be to search for something that bears some resemblance to something you know to be edible. You will probably avoid things that look or smell rotten (such as durian). You may prefer items that are brightly colored, because that means ripeness, which usually means sweet and nutritious. You may also prefer bright green broad-leafed plants, or tender stalks and shoots. Your meat choices will probably lean towards small mammals, and perhaps insects such as termites. Countless accounts of people stuck in the wild tell the same story: When in doubt, we will tend to gravitate towards the types of foods we've eaten throughout our evolutionary history, while eschewing other possibilities unless we have no other choice.

Having examined and smelled our prospective meal for apparent freshness and ripeness, the next step is to take a bite—a very small bite—and evaluate for flavor. Bitterness usually signals poison, so we'll reject those items. An item that passes all these tests will be chewed and swallowed, after which we'll wait a while (probably several hours or longer) to see what happens. Chances are excellent that any poison will reveal itself within this time frame, and that tiny bite will only impart a manageable dose. If there are no ill effects, we will take a larger bite and wait once more. This cycle will repeat itself until we either encounter a problem, or eat so much as to be confident of our new food source. It is entirely possible that we'll consume a slow poison or one that requires a large dose to take noticeable effect, but this risk is minimal compared to the imminent risk of starvation. This is how animals try new foods, and how every human survival manual tells us how to do it. This is one of nature's biggest defenses against poison.

This system works so well, because a poisonous plant or animal doesn't want to be eaten. It makes little sense for that plant or animal to evolve a slow poison, because that gives a hungry experimenter more time to do more damage. Over time, evolution would tend to favor those individuals with fast-acting poisons, because more of them would remain around to reproduce.

Finicky Children

The child who will only eat skinless chicken meat, white rice, and broccoli is exercising all due evolutionary caution. She or he has selected proteins, grains, and vegetables that s/he knows are good to eat, and has little reason to stray from those preferences until s/he is old enough to withstand the potential risks of branching out. Strong flavors such as spice, and textures such as sauces, can trigger protective instincts. The child isn't trying to be difficult; it is trying to avoid poisoning. Bland flavors and simple textures spell security, especially to a young nervous system that is practically overwhelmed with new stimuli every day.

Our minds are like our stomachs; they are whetted by the change of their food, and variety supplies both with fresh appetites.
Quintilian

This is just one example. Children are most likely to eat what their parents eat. This begins during breast feeding. Chemicals from food pass to the child in mother's milk, which predisposes the child to like foods containing those chemicals. If you are a nursing mother who eats a lot of chicken prepared in all manner of styles, then you are passing on chicken chemicals to your baby. This child will probably develop a strong taste for chicken, and possibly even a preference for plain chicken, because plain chicken lacks any means of masking the chemicals the child seeks through spices, sauces, etc.

This is why it's critically important for mothers to eat a healthy diet rich in fruits and vegetables while pregnant, while nursing, and thereafter. The child will mimic its parents' diets, because it lacks any experience of its own, and will follow the only cues it has. The child cannot do anything else because evolutionary rules dictate that straying from the tried and true increase the risk of poisoning. Logan's mother and I ate plenty of vegetables and fruits, and more than one person has described him as the only kid they know who goes around asking for cauliflower, broccoli, carrots, lettuce, tomatoes, and other veggies at dinner parties. I couldn't be happier.

Follow your dreams, work hard, practice and persevere. Make sure you eat a variety of foods, get plenty of exercise and maintain a healthy lifestyle.
Sasha Cohen

When Things Go Wrong

Poisons are an iffy way to catch rats, because they have a long memory. If a dose of poison fails to kill a rat, it will remember that experience for the rest of its life. Humans have a similar defense mechanism built into our own psyches. Our first line of defense needs no prior experience to be active: We eschew foods that we may intellectually know are perfectly safe, because they just don't fit our individual preferences. If you've watched the television show *Iron Chef* (the original Japanese version), then you have an idea of the amazing breadth of things our species can eat. How much of that breadth you're open to trying for yourself is a very different proposition. I can only imagine that some foods were first eaten on a dare. We fear unfamiliar foods, and rightly so.

Our second line of defense is to avoid food that has made you sick in the past, especially if you got sick the first time you ate that food. If you're stranded somewhere and the first bite of something unfamiliar makes you sick, you won't eat it again. This usually holds true even if you get sick the first time you eat something you know to be edible (such as if your first experience with oysters goes horribly wrong). We may even swear off foods we've previously eaten successfully, should we have the misfortune to ingest a bad batch. We are most forgiving of foods we have extensive experience with. For example, I've never gotten sick from eating chicken, and will almost certainly keep eating chicken should that ever change.

This brings up another important part of our defense against poison, namely the ability to discern whether something is inherently poisonous (that leaf on the deserted island), or whether we just happened to eat a bad example of something that is normally good for us (a tainted chicken breast). Excluding a normally good food just because we got a bad batch risks cutting off a valuable food supply, resulting in a classic case of, "water, water everywhere, and not a drop to drink."

The Dietary Conundrum

The staggering array of possible foodstuffs, plus the attendant risk of poisoning, combined with the need to allow some exceptions, creates a complicated set of rules that must be followed faithfully, yet not too absolutely. As I mentioned above, this requires some equally complicated mental circuitry in order to handle these many conflicting demands. This creates a true dilemma. Ancient humans faced this dilemma as they migrated across the planet encountering new species everywhere they went. Modern humans face this dilemma, thanks to a global economy that distributes food from just about everywhere to just about everywhere. Our food production and distribution systems pose dilemmas of their own, not least because they are a significant source of the habitat change and pollution taking place today. That we have navigated these waters as well as we have is a testament to the power of the human intellect.

I've been on a diet for two weeks and all I've lost is two weeks.
Totie Fields

Gourmets vs. Gourmands

An animal with a large supply of stationary, brightly-colored fruit has the luxury of being able to eat at leisure. Find enough trees that fruit at different times of the year, and life becomes a never-ending buffet that meanders from tree to fruit-laden tree. The fruit isn't going anywhere, nor will a few hours or even a few days make much difference in most cases. This encourages slow snacking throughout the day. Why set aside long periods of time to eat large quantities when you can't carry much while climbing trees? Why rush to take on as much food as possible when the next bite is wating patiently only an arm's length away? Monkeys and apes know nothing about breakfast, lunch, and dinner, nor do they have any reason to hurry up and eat. Not surprisingly, they eat many small snacks throughout the day, and take the time to savor their

A gourmet who thinks of calories is like a tart who looks at her watch.
James Beard

Chapter 15
The Omnivore's Dilemma

food. Our evolutionary cousins are the gourmets of the animal kingdom.

A hunter animal has no such luxury. Its prey doesn't want to be eaten, and will scatter at the first sign of trouble. Finding food requires a lot of patience, and catching food requires an energy-draining sprint. Each failed sprint leaves the predator with fewer reserves for a following try while its prey can simply load up on more grasses, fruits, or vegetables at any time. Problems multiply once the prey is killed, because other animals will try to steal some or all of it. Meals are few and far between for a predator, who must take therefore advantage of every opportunity to eat as much as possible as quickly as possible, lest another steal the meal. Predators are the gluttons of the animal kingdom by necessity.

An animal that evolves from eating fruits to deriving a large portion of its calories from hunting could be expected to eat in a manner somewhere between the two. This animal might eat frequently, but in set meals instead of ongoing snacks. It might also take the time to enjoy its food, while not devoting inordinately long times to feeding. This situation describes the human condition perfectly. Most of us eat three main meals per day, with perhaps a snack or two on top of that. We enjoy being able to enjoy our food, and go so far as to adorn it with herbs, spices, sauces, and other embellishments, whose only purpose is to enhance the dining experience. We are inherently able to snack, and some of us do eat that way, but most people stick to the "three squares per day" routine, even the vegetarians and vegans among us. Our daily routines probably evolved in response to our hunter-gatherer lifestyle, in which the hunters ate before setting out and again when they returned. Seen against the two extremes of our monkey cousins and predator animals, humans are the gourmands of the animal kingdom.

Other Eating

There is nothing quite as satisfying as a full belly and the sense of well-being that comes just after eating a good meal. It is therefore no surprise that some people, myself included, eat when they are not hungry, eat more than they need to fill themselves, and/or eat foods that may taste good but that lack nutritional value. We eat to alleviate stress, to redirect anger or

---

Red meat is not bad for you. Now blue-green meat, that's bad for you!
Tommy Smothers

If only it was as easy to banish hunger by rubbing the belly as it is to masturbate.
Diogenes

other emotions, and for many other reasons that have nothing to do with staying alive. Such eating is called *displacement eating*. Chewing gum is a prime example of displacement eating.

Vegetarianism?

A sizeable number of humans choose not to eat meat. Those who avoid all animal products are called *vegans*. There are many reasons why people may choose to restrict or even eliminate meat and/or animal products, such as eggs, butter, milk, cheese, etc. from their diets. Some people claim that humans are not designed to eat meat. We evolved from a common ancestor of humans and chimpanzees. Chimpanzees eat meat. In boom times, a chimp's diet will have about as much meat as a human diet during bad times. There is every reason to assume that our common ancestor species ate at least some meat. Humans are most certainly designed to eat meat.

Some people abhor the cruelty of killing animals for food. I certainly don't enjoy making animals suffer, but the simple fact is that humans are the only ones concerned about the humane treatment of animals. Predator animals don't concern themselves with providing their meals a swift, painless death. On the contrary, most prey animals must endure a chase ending with a slow death and possibly even being eaten alive. There is nothing genteel about a male praying mantis being decapitated and eaten by its mate.

The idea that vegetarianism is kinder or more natural than eating meat is therefore completely bankrupt in my opinion. Still, there are some compelling arguments for at least examining how we raise our food, and some equally compelling reasons to consider making some sweeping changes.

Modern farmed food is a far cry from its wild counterparts. Prey animals may suffer at the end of their lives, but at least get to live as nature intended until then. They are not confined in excessive numbers and densities, and made to eat unnatural foods in modern farms. The treatment all too many farm animals receive is abhorrent and actually elevates the chances of a truly deadly epidemic that could sicken or kill many thousands of people. Most of the meat people eat today was raised under

Vegetarianism is harmless enough, though it is apt to fill a man with wind and self-righteousness.
Sir Robert Hutchinson

Most vegetarians look so much like the food they eat that they can be classified as cannibals.
Finley Peter Dunne

Chapter 15
The Omnivore's Dilemma

conditions that lessen its nutritional value, while also greatly increasing its fat content.

The sheer number of people on this planet and the disproportionate amounts of land and food required to raise animals is another compelling argument for at least reducing our meat intake. We could feed many people using the same resources required to raise one cow while simultaneously reducing the toxins (fertilizers, etc.) that concentrate inside the meat only to pass into our own bodies. We could reduce—if not eliminate—world starvation by reducing or eliminating meat from our diets.

In short, some of the most compelling reasons for adopting a vegetarian lifestyle have nothing to do with nature itself, and everything to do with how humans have manipulated that nature. Chapter 16 explores how agriculture and modern factory farming combined with mass marketing campaigns have created a system of eating that is causing human suffering on an almost unimaginable scale.

Nothing will benefit human health and increase the chances for survival of life on Earth as much as the evolution to a vegetarian diet.
Albert Einstein

Entitlement

If we can do something, why shouldn't we? There is no naturally evolved trait that its owner fails to use to full advantage. Life exists for the purpose of passing on genetic material to succeeding generations, and the only way to do that is by winning a round in an endless competition for resources. If a human anywhere on this planet is able to enjoy food from all over the planet at any time of year, why shouldn't s/he do so? It is December in San Francisco as I type this, but I can eat amazingly delicious cherries thanks to a huge fleet of ships and airplanes that have made growing seasons almost a thing of the past. If the cities on America's east coast are too large to live from locally grown produce, why not ship it in from places like California?

Humans have elevated this question and the resulting sense of entitlement to an art form. The average calorie in the United States travels well over 1,000 miles from production (farm) to consumer. Some foods, like the winter cherries in San Francisco, have traveled from the other side of the world (such as Chile) to reach local fruit stands. I won't lie; I am a sucker for

To learn to get along without, to realize that what the world is going to demand of us may be a good deal more important than what we are entitled to demand of it—this is a hard lesson.
Bruce Catton

good cherries, and being able to enjoy them twice a year makes me very happy indeed. But at what point does the mere ability to eat almost anything at any time from anywhere become a liability? This is a moral question, and I'm reserving most of the moral questions in this book until the very end, but I have to wonder: Given the state of the world and where that state could lead us, should we be reexamining some of our most fundamental ideas about how we grow, process, ship, and eat our food? Thanks to modern technology, this question may well lie at the heart of a modern omnivore's dilemma.

I have often thought morality may perhaps consist solely in the courage of making a choice.
Leon Blum

Chapter 16

What's for Dinner?

The food most of us eat today may outwardly resemble its wild cousins, but that's where most of the similarity ends. Thousands of years of agriculture have yielded entirely new strains of ancient foods, and modern farming and processing methods have created results beyond the mythical Dr. Frankenstein's wildest imagination. Today, the food we eat and how we eat it has the potential to do far more harm than good. Two identical people ostensibly eating the same diet could experience widely different life spans, with equally different health patterns. There is no such thing as a free lunch, and the very means we employ to provide lunch to as many people as possible have lessened the value of each individual lunch.

Agriculture

A farm is an irregular patch of nettles bounded by short-term notes, containing a fool and his wife who didn't know enough to stay in the city.
S.J. Perelman

Picture yourself as a primitive hunter-gatherer with no knowledge of agriculture or farming. Life is tough. You spend most of your days searching for food that must then be hauled some distance back to the home for processing. A bad year can have you wandering even further and getting even less for your efforts than usual. One fine day, you and your tribe crest a hill and peer down into the next valley, where you discover rows upon rows of fruiting trees. Next to this astonishing sight is an entire field full of edible grains. If that wasn't enough, an entire herd of animals is milling about inside a wooden enclosure. Just how many milliseconds do you think it would take for you to grasp the tremendous advantages of agriculture over wandering in search of food every day? Why then has agriculture spread so slowly during its 10,000-year history? Why do some tribes refuse to abandon their hunting and gathering ways to this day?

Roots

Primitive humans foraged for wild plants of all types, and brought their bounty to the tribe's settlement. Life was more or less nomadic as people wandered in search of sustenance, but rich pickings could have led to fairly long stints at the same location. It was only a matter of time before people noticed that the fruits, vegetables, and grasses they were eating were beginning to grow in and around their homes. From there, it only required a moderate mental leap to associate seeds discarded during food preparation or scattered in excrement with the appearance of these food plants. The ever inventive human intellect took over, and soon our ancestors were scattering seeds deliberately, and reaping the rewards of their efforts. Further trial and error led to developments, such as irrigation, fertilization, plowing, and eventually the modern factory farm that churns out more crops per acre than the first farmers could ever have dreamed possible. Along the way, happy accidents and deliberate experiments gave rise to new strains and hybrids of ancient crops. For example, modern corn is a descendant of teosinte, a barely edible grass that is hardly recognizable as anything worth cultivating. Today, genetic engineering is yielding new crops far more quickly and efficiently than any selective breeding program could ever hope for.

Many of the new primitives regard the beginnings of agriculture as one of humanity's major steps in the wrong direction.
Walter Truett Anderson

Property

We have already seen that humans are highly territorial animals. Our ancestors claimed and defended territories against rival groups. Agriculture exacerbated these natural tendencies. Farming is labor intensive work, and the farmers needed some assurances that their crops would be secure against theft by other tribe members or pillaging by outside invaders. Subdividing a tribe's territory into individually owned patches of

Women constitute half the world's population, perform nearly two-thirds of its work hours, receive one-tenth of the world's income and own less than one-hundredth of the world's property.
United Nations

land solved both problems. Tribes adopted penalties for trespassing and theft, giving rise to the concept of property rights. The presence of a steady food supply required a steady defense, which gave rise to soldiers and standing armies.

Specialization

The test of a vocation is the love of the drudgery it involves.
Logan Pearsall Smith

Farming requires only a relatively small percentage of a group's members devoted to supplying food for everyone. This leaves everyone else with a lot of time on their hands, time that could be—and was—put to other uses. Some people had to grow the plants and animals for food (farming). Others had to defend this territory against invaders (soldiers). Still others were needed to construct shelters, make tools, enforce laws, survey land, and administer the whole enterprise. In short, the advent of agriculture allowed—and even demanded—increasing specialization of skills and services. This specialization even extended to those directly involved in agriculture. Some might be shepherds, while others grow specific crops.

The seemingly simple achievement of being able to guarantee a relatively stable food supply had ripple effects that forever altered the lives of those tribes that opted into the agricultural revolution. Agriculture led to civilization as we know it. So why was it so slow to catch on?

Restriction

Variety is the soul of pleasure.
Aphra Behn

A foraging human has a wide array of potential food choices, thanks to our omnivorous digestive systems. Even a relatively finicky forager still has plenty of nutritional possibilities, because the same species of plant or animal growing in two separate locations may have different levels of different nutrients depending on soil conditions, lighting, moisture levels, and other factors.

Agriculturally produced food lacks this nutritional variety, because the crop grows in the same soil season after season. At some point, the soil's nutrients will be used up, at which point the farmer must either let the field go fallow for some time (unproductive), rotate crops that use different nutrients (a viable alternative that can still require the field to go fallow for some time during the cycle), and/or resort to artificial fertilizer (starting, presumably, with animal dung).

Further complicating matters, most people in a tribe will probably agree that certain foods are more desirable than others because of taste, ease of preparation, and (not least of all) ease of growing. Farmers grow this desirable food in large quantities and marginalize other crops. This further restricts nutritional variety. Burgeoning populations (any animal will reproduce more when food supplies are good) require ever more food, forcing farmers to find ways to boost yields. The vicious cycle continues. Achieving efficiencies of scale demands large-scale operations with relatively few crops, further reducing nutritional variety. Today, more than 50% of the world's food calories originate in wheat, rice, and corn. That steak you ate for dinner probably came from a corn-fed cow, a fact that might not seem noteworthy, but for the fact that cows evolved to eat grass. Why grow grass that has only one, use when there are dozens of raw products and thousands of foods that can be made using corn? Corn has even replaced sugar as the sweetener of choice in many food products. Next time you eat something sweet, see if that sweetness comes from high fructose corn syrup. Cane and beet sugars have more nutritional value than corn syrup.

Tell me what you eat, and I will tell you what you are.
Anthelme Brillat-Savarin

The combined effects of soil depletion and increased reliance on fewer crop species has severely curtailed the nutritional variety available to many people today. I have yet to discuss disease, the deliberate poisoning of our food supply, or the many ways in which food is processed. Each of these factors further diminishes the value of the food we eat today compared to that eaten by our "primitive" ancestors.

Nutritional restriction isn't the only problem agriculture brings. Social inequality, disease, and shortened life spans are some of the other by-products of growing our own food.

Inequality

A hunter-gatherer is free to roam her or his tribe's lands, and has a reasonably equal claim to available resources. There is no individual land ownership, because that could threaten the entire tribe's survival. Agricultural societies benefit from individual land ownership, because this gives farmers incentive to

That's part of American greatness, is discrimination. Yes, sir. Inequality, I think, breeds freedom and gives a man opportunity.
S.J. Perelman

work their land to provide for everyone. I certainly wouldn't tend a crop that anyone could help themselves to at any time. Farmers need to be assured that their crops are secure. Anyone who steals crops threatens both the farmer's livelihood and the tribe's survival. Laws and penalties are therefore required, as are people to enforce them. Agricultural lands are easy pickings for invaders and require standing armies to protect them. As I noted above, this leads to increased specialization of skills and professions. I am sitting in front of an electronic device pushing keys that are transcribing thoughts into words that I intend to sell—an activity about as far removed from hunting, gathering, and foraging as I can imagine, yet one designed (in part) to ensure my own food supply.

> *The world makes up for all its follies and injustices by being damnably sentimental.*
> Thomas H. Huxley

Where does this leave people who don't own land and who are not part of the power elite? Enter the concept of working for others, where one person profits from the labors of others. Anyone with a job is earning far more for their employer (and the employer's investors and shareholders) than they are for themselves. One good friend of mine earns a 2% commission on the jewelry he sells, making his work 98% inefficient. It is certainly true that his employer incurs significant costs for raw materials, utilities, taxes, insurance, etc. Still, 2% is 2%.

The disparity grows when we factor in the relative value of different kinds of work. Most anyone can dig a ditch, so ditch-digging is cheaply paid work. Relatively few people have the skills and qualifications to program computers, so software engineers tend to be relatively highly paid. A loaf of bread costs the same no matter who buys it. The ditch-digger who may need access to the best food available in order to maintain her or his health in the face of backbreaking work suddenly becomes the person least able to afford that food. The software engineer who sits in a chair and hardly moves all day may have the least nutritional needs, but can afford the best foods. This paradox and others like it are all possible thanks to agriculture.

> *In the part of this universe that we know there is great injustice, and often the good suffer, and often the wicked prosper, and one hardly knows which of those is the more annoying.*
> Bertrand Russell

If farmers can own land, why not others? There is only a finite amount of land available, meaning that not everyone can own a piece of the Earth. Those who don't own land must pay rent to live and work on other people's land. Those who don't own land are the ones in the lower tiers of the pay scale. Land ownership and landlording has been a tried and true investment for as long as people have been able to own land. The invention of wills and inheritances allowed land and property to pass from generation to generation within families. One could enjoy significant power and luxury simply by having the good fortune to be born into the right family. Today, a woman named after a building (the Hilton hotel in Paris, France) has far more power than most other people, simply because she was born into the right family. A construction worker helping build one of her family's hotels will produce far more of actual value than the woman herself ever will, but won't enjoy any of the rewards. How can this be?

Different professions that pay different wages create inequality of wealth and status. Mechanisms to preserve and concentrate that wealth (inheritances, employment, etc.) exacerbate that inequality to the levels described in Chapter 11.

Hunter-gatherers have none of this. They reap all of the fruits of their labors. Anyone who finds food is free to eat her or his fill, and may opt to give some away as a hedge against future bad luck. It is a 100% efficient operation that lacks both the specialization and inequality of agriculture. The group's fortunes wax and wane with the hunt. These people don't care what building you're named after. You can either pull your own weight, or you can't. It's an eminently fair system.

Though the vicious can sometimes pour affliction upon the good, their power is transient and their punishment certain; and that innocence, though oppressed by injustice, shall, supported by patience, finally triumph over misfortune!.
Ann Radcliffe

Food Poisoning

Inject a large supply of food into an environment, and species that feed on that food will experience a population boom. That law of nature works for any species imaginable. Imagine a species of aphid that feeds on a certain species of grass. Now imagine an aphid discovering a huge field of its favorite meal just waiting to be eaten. Do you think this aphid's numbers will explode? Now imagine fields of this grass getting bigger and bigger, and ever more densely packed with ever more bountiful varieties of the same grass. This aphid species is set for a tremendous expansion.

In agriculture, large fields are better than small fields, because farmers can benefit from economies of scale. Farmers also benefit from harvesting as much as possible from their fields. Throw in the restriction of crops to a relative few (remember that over half the world's calories come from just three crops)

When you consider what a chance women have to poison their husbands, it's a wonder there isn't more of it done.
Kin Hubbard

and you have a *monoculture*, vast tracts of lands occupied by the same crop. Huge fields of wheat (the grass I mentioned above) present huge opportunities for aphids and other species that enjoy eating wheat. The resulting population explosion can threaten the entire wheat harvest. Small, widely scattered fields of wheat would minimize the pest problem, while preventing the farmer from leveraging the benefits of large-scale operations. Thankfully, the farmer has a secret weapon, one that can prevent pest infestations while allowing large-scale agriculture to occur: poison. Farming uses almost three pounds of pesticide per acre. Pesticide is not nice stuff. On the contrary, it's designed to kill. Methyl bromide, a pesticide used on strawberries, is so toxic that it renders soil virtually sterile.

Even worse, it's all but guaranteed that at least a few pests will survive a pesticide application. These survivors are now free to eat their fill without any competition from their fallen brethren. Moreover, they are resistant to the pesticide in question. Over time, the pesticide becomes less and less effective. This forces the farmer to keep seeking newer, more toxic chemicals to keep the pests at bay.

Anything that can be done chemically can be done by other means.
William S. Burroughs

The problem with pesticides is that they kill indiscriminately. Chemicals have no way to tell a pest from a beneficial earthworm from a human. It's also impossible to totally remove pesticides from the plants we eat (remember that they have to withstand rainfall and irrigation) or from the animals that eat those plants and concentrate the poisons in their flesh before being slaughtered for meat.

Stop using pesticides overnight, and our food supplies will vanish under swarms of hungry pests. Keep using pesticides, and we'll be able to feed billions of people, just so long as we don't mind poisoning them a little. Farmers worldwide are working on ways to reduce or even eliminate pesticides, an effort I applaud wholeheartedly.

Changing Diets

A modern cow and its wild ancestor may belong to the same species, but that's where much of the similarity ends. The wild cow moved about freely, grazing on an abundant diversity of grasses and small plants. This varied nutrition concentrated in the cow's tissues, which were lean and robust from constant motion. The cow's hooves trampled and broke the dirt underneath it, aerating the soil and helping the grasses it ate to grow. Its meat was tough and hard for a person to chew because of its leanness. It was also very healthy for that person to eat, being both fat-free and rich in heart-friendly omega-3 acids and other proteins.

The modern cow is forced to eat corn (which stresses the animal and makes it vulnerable to disease), which lessens the meat's nutritional value and, combined with lack of exercise, vastly increases the *marbling* (tissue fattiness). This meat is tender, juicy, flavorful, commands a high price at fancy restaurants, and is laden with omega-6 acids that are extremely bad for the heart, colon, and other organs. The red meat eaten by most people today thus contributes to heart disease, colon cancer, and other ills. The red meat eaten by people in the Mediterranean region has none of these ill effects, because those animals feed on natural pastures.

Agricultural plants tend to be high in *carbohydrates* (sugar) and lacking in vitamin and mineral nutrients. They also tend to be laden with pesticides, as discussed above.

You may be thinking that agriculture with its drawbacks remains infinitely preferable to hunting and gathering. Consider that the average African Bushman eats over 2,000 calories per day, which includes 93 grams of protein. Their diet includes more than 85 types of plants. Disease is rare, and famines are all but unknown. Not bad for a "primitive" society! (Information from *The Omnivore's Dilemma* by Michael Pollan.)

Archeological evidence indicates that meat intake has risen dramatically over the last few thousand years. Beef intake has certainly spiked in the last half-century, from a relative luxury to a daily staple. We are now eating far more meat than ever before, and this meat is increasingly laden with unhealthy fats. This is just one of the ways in which modern food processing is altering our ancestral eating habits.

Fat

Fat is a ready source of quick energy. Most if not all animals have at least a little fat, which serves as both insulation and

I don't even butter my bread. I consider that cooking.
Katherine Cebrian

I am resolved to grow fat, and look young till forty.
John Dryden

emergency food reserve. Fat can be easily digested and provides a large caloric punch in a small package. I've already mentioned that wild animals tend to be low in fat because they are constantly exercising. There are notable exceptions, such as whales, mammoths, and other animals that have large fat reserves in order to survive in cold environments. The people who ate these animals were just as cold and needed the calories to stay warm. Those exceptions aside, most people ate very lean meat—too lean for most modern people to consider palatable. Fat was therefore a relatively rare—and highly valuable—treat.

Modern food production methods leverage our preference for fat to the hilt. Meat is raised so as to maximize marbling. Butter, cream, oils, greases, and more find their ways into many of the foods we eat, particularly processed and prepared foods. Check the labels on the foods you eat. The percentage of calories from fat may astonish you. Why do food processors add fat to food? Because fat tastes good. We evolved our taste for fat in a time when the scarcity of fat made it a precious commodity. This is one more example of technology advancing far too rapidly for our brains to keep up.

Sugar

Sham Harga had run a successful eatery for many years by always smiling, never extending credit, and realizing that most of his customers wanted meals properly balanced between the four food groups: sugar, starch, grease, and burnt crunchy bits.
Terry Pratchett

Sugar (carbohydrates) provides an extremely rapid burst of energy. Fruits are naturally rich in sugar, because they literally want to be eaten. The fruit benefits from having its seeds scattered to pass along its genetic material, and the animal lucky enough to eat that fruit benefits from the aforementioned energy rush.

Modern food producers take advantage of humanity's natural sweet tooth in many ways. Our three chief food crops (wheat, rice, and corn) are all rich in carbohydrates. To this, we add hefty amounts of sweetening agents, such as sugar from sugarcane or beets, or—increasingly—high fructose corn syrup, a nutritionally bereft substance whose only claim to fame is its sweet flavor and a fleeting rush of energy. Check those food labels! The amount of sugars, natural and otherwise, in what you eat may surprise you. Sugar tastes good. Good-tasting food sells better than food that doesn't taste good.

Salt

Salt is a critical electrolyte. A contestant on a radio show once died from drinking too much water, which completely upset her electrolyte balance. Salt is rare in nature. Some salt is found in meat, but hunter-gatherers tended to gather far more than they hunted. This made salt a rare and valuable commodity. Salt licks and other natural sources of salt were highly prized. To this day, camel caravans haul salt across the Sahara desert, and bridal dowries are paid in salt.

Modern salt harvesting and mining techniques have made this commodity cheaper than dirt, and food processors have not been idle. Check the salt content of the food you eat. I once saw a small microwaveable burrito (yes, I have a weakness for those) that contained over 60% of the recommended daily allowance of sodium. Salt tastes good, and good-tasting food outsells those that don't taste so good.

Give neither advice nor salt, until you are asked for it.
English Proverb

Ancient Palates

The human palate, like the human brain, is still roaming the ancient African savanna. Modern food processors cater to this palate by artificially boosting the amounts of fat, sugar, and salt in our food beyond nature's wildest expectations. Small wonder that getting children to eat plain vegetables can be so difficult! Convincing a young brain that thinks it's in the middle of ancient Africa that a bland food has more benefits than the tasty snack is all but impossible. Today's technology has proven adept at getting people to eat way too much food, food that has been deliberately tampered with to appeal to our ancient palates.

The mismatch hypothesis proposes that many of today's most pressing health issues (such as obesity, heart disease, cancer, and others) are a direct result of the mismatch between the characteristics humans evolved with and our present environment. Given how thoroughly we have altered our food supply, this is not terribly difficult to imagine.

You can tell a lot about a fellow's character by his way of eating jellybeans.
Ronald Reagan

At All Costs

Why not go out on a limb? Isn't that where the fruit is?.
Frank Scully

Be fruitful and multiply. That is nature's imperative, to pass along the genes at all cost. Humans have taken this biological mandate and run with it. The world's population recently topped 7,000,000,000 (seven billion) people. Humans have mastered the art of quantity.

But what about quality? Agriculture has brought disease, shorter life spans, famine, and extreme social inequality to humanity. Those of us who can afford to eat, ingest foods that have been engineered and altered almost beyond recognition. Some people, those in the elite, benefited tremendously from the advent of agriculture and civilization. Most people lost at least some of their freedom, their varied natural diet, and even their ability to fend for themselves. Population density in agricultural societies is 10 times that of hunter-gatherer societies, and the former reproduce twice as much as the latter. Agriculture consumes the majority of the world's available fresh water supply, and causes pollution and environmental degradation.

It is easy to imagine the fictional tribe I mentioned at the beginning of this chapter embracing agriculture at first sight. It is equally easy to imagine that many hunter-gatherer societies observed agriculture's profound effects and decided to opt out of the whole exercise. The mere presence of hunter-gatherer societies in today's modern world should prove that agriculture is a mixed blessing, at best.

Reason for Hope

My doctor gave me two weeks to live. I hope they're in August.
Ronnie Shakes

I must in fairness point out that my mere presence on this Earth is probably due to agriculture, as is yours. On that selfish level, agriculture has been extremely beneficial. There are also plenty of reasons to hope for a brighter agricultural future. Here are just a few:

- Awareness of how pesticides affect humans and the environment is spreading, as are alternatives such as organic farming and other methods of pest control.

- There is a growing interest in exotic crops, and several efforts are underway to preserve seeds of rare crop species.

- Emerging technologies are allowing farmers to save water and reduce fertilizer usage.
- Innovations such as vertical farming (hydroponic farms in skyscrapers) could yield a wide variety of organic crops grown where they are consumed, thus reducing both shipping time and resources required.

Our imagination is the only limit to what we can hope to have in the future.
Charles F. Kettering

Chapter 17

Bringing Home the Bacon

This chapter presents some additional interesting information related to what we eat, and how the need to find food has shaped our brains. Hunter-gatherers tended to lean far more toward the gathering end of the spectrum. Big-game hunting tended to be very inefficient, and entire species vanished as a result. Next, the gambler's fallacy, confirmation bias, and the pleasure derived from seeking to fulfill our needs are all part of a long evolutionary legacy designed to help us survive. Finally, this chapter concludes with a few more thoughts on population quantity versus quality.

Bagging Dinner

Hope is a good breakfast, but it is a bad supper.
Sir Francis Bacon

As we saw in Chapter 7, predation (hunting) is a complex process that requires careful planning, flawless execution, and perhaps a mistake or two on the prey's part if the hunter hopes to make the kill. Hunting requires a significant investment of time spent making weapons, trekking to hunting grounds, searching for prey, and lying in wait. It also requires a great deal of energy, particularly during the final chase/battle.

Consider a rocket such as the Saturn V used to send astronauts to the Moon. The entire rocket weighed in at around 6,200,000 pounds (3,100 tons), while the payload consisting of the Apollo capsule and service module weighed a mere 68,000 pounds (34 tons). 99% of the total weight was consumed by the rocket engines and fuel required to boost the payload to lunar orbit. This disparity occurs because the rocket needs enough power and fuel to send the payload to its destination. It then needs fuel and power to lift the fuel and power that will send the payload to its destination, then more fuel and power to lift that fuel and power, and so on. The same principle applies to large expeditions, where many dozens of porters and climbers manage to send a small team to the summit. The 1963 American expedition to Mount Everest required about 900 porters, and placed six people atop the mountain.

Hunting is similar, in that every activity related to hunting consumes precious calories that, once expended, are gone forever. Involving more people in a hunt may increase the odds of bagging a large animal, but also increases the risk. A hunter who fails to make a kill has fewer resources with which to try again. It may be more efficient to forage for plants, since plants are readily found and won't run away when approached. Scavengers are also far less likely to try to steal plants than freshly killed animals. All animals, humans included, must be very careful about choosing where, when, what, and how to hunt in order to risk wasting as few calories as possible.

Popular legend, and even many tribal myths, glorify the hunt and give the impression that our ancestors consumed meat in quantities that would make a modern dietician die of shock. But was this really the case? The huge caloric investment and high risk of failure inherent in hunting seem to argue against meat as a huge factor in ancient human diets.

Why then make such a big deal about hunting? A successful hunt was a boon to a hunter-gatherer tribe because it offered a huge supply of calories rich in protein, hides for clothing and shelter, bones for tools and weapons, and more. Lessons from modern news media make it safe to assume that such occasions were the exception. Plane crashes, even those involving small planes, are big news. Why? Because they are exceedingly rare events. If every day saw airliners dropping from the skies like flies, then newspapers would soon cease reporting all but the gravest incidents. For example, few news papers report traffic accidents, unless the accident in question is exceptional in some way. Using the same logic, the mere fact that ancient

Nobody realizes that some people expend tremendous energy merely to be normal.
Albert Camus

In this life we get only those things for which we hunt, for which we strive, and for which we are willing to sacrifice.
George Matthew Adams

humans made such a fuss about hunting probably means that hunts—particularly successful ones—were a relatively rare occurrence. This rarity combined with the large benefits that did come from success make it easy to imagine an extremely grateful populace that would devise ceremonies, repeat the tale as often as anyone cared to listen, and even worship the animals themselves.

Does this theoretical thinking pan out in the real world? According to Jared Diamond in *The Third Chimpanzee*, most New Guinea hunters will, when pressed, admit to killing only a few large animals in their entire lives. Further evidence comes from looking at chimpanzees, who survive by eating plants and the occasional small animal. Big-game hunting did not emerge until humans had evolved modern anatomy and reasonably modern behaviors, including the ability to make weapons. And what weapons! Archaeologist Julian H. Steward proposed the radical idea that stone "scrapers" were actually designed to be thrown at prey. Subsequent tests with different so-called scrapers have lent strong credence to the idea that many of these supposedly primitive handheld tools were in fact sophisticated ranged weapons. One of these rocks hitting an animal's spine could cause the animal to reflexively fall over, giving hunters a crucial window of opportunity to close in and finish it off.

Hunting was the labour of the savages of North America, but the amusement of the gentlemen of England.
Samual Johnson

The archeological record seems clear: Humans have been fairly inefficient hunters for most of our evolutionary history, at least until the development of firearms and other accurate ranged weapons and other technology. Hunting under primitive conditions may have been a risky undertaking, but hunting on horseback with a bow or gun at hand (possibly with other hunters and/or dogs to help) is a completely different proposition, one that elevated—or downgraded—hunting from survival tool to sport.

Apropos of big game, the archeological record indicates that humans quickly wiped out most of the available large animals. Mammoths went extinct, as did plenty of other species around the planet such as the mastodons, lions, camels, and others that used to wander the North American continent. Early Americans hunted big game by herding them to mass deaths off cliffs or into swamps to be shot with poisoned darts. These massive kills were probably safer than trying to take on individual animals whose companions might come to the rescue. (Elephants are known to aid other elephants in distress, for example.) They probably also provided the hunters with far more meat than could ever be shared among the entire tribe before it spoiled. Over time, the largest prey animals vanished, and smaller animals became the staple foods of humans and predator animals alike. The relative timidity of deer and the manageable size of their carcasses may well have contributed to the deer's ongoing survival.

On the Hunt

If our omnivorous diet has contributed to the sophistication of our brains, then hunting and gathering may have ingrained some mental programming that has had some pretty interesting side effects, such as superstition and the extreme profitability of gambling casinos.

The Gambler's Fallacy

The so-called *gambler's fallacy i*s a belief that the outcome of any single event in a series of random events depends on the outcome of previous events in the series. For example, someone watching a roulette wheel may believe that a black number is increasingly likely to occur because a long string of red numbers has been occurring. Casinos encourage this belief by posting signs displaying the wheel's last several results, and gamblers invariably respond by betting opposite the current streak. The problem with betting like this is that the odds of any number occurring on the next roll of the wheel are exactly the same as they've always been. If a black number has a 50% chance of appearing, then it has the same 50% chance no matter how many times red numbers appear in a row. There is no more or less advantage to placing one's bet on a red or black number, because either outcome is equally likely. Basing one's belief about a future event based on past events and not the game's actual odds is a losing proposition for the gambler, which is why casinos display those signs. They want you to lose and lose big.

An expert is a person who avoids small error as he sweeps on to the grand fallacy.
Benjamin Stolberg

Here's the catch: Casinos don't have the advantage over players because they are unfair. In fact, players would quickly spot and take advantage of any unfair wheel, loaded dice, or similar situation, to the house's great disadvantage. This is why so many people try to beat the house by cheating—by making the system unfair. Casinos win precisely because they are perfectly fair. They create situations where red and black numbers have perfectly even odds of appearing (among too many examples to list). Past behavior is no indicator whatsoever of future behavior when it comes to gambling (or for any so-called random event).

Nature is inherently unfair in that past performance is a great indicator of future performance. The sun has been rising after a period of darkness for as long as each of us has been alive, and we therefore have every reason to think the sun will come up tomorrow. One could be forgiven for thinking that the next day will be warm based on today's sunshine, cloudless sky, and calm wind. One could also be forgiven for looking for food in areas that have previously shown promise. After all, if it's dry and animals clustered around the water hole yesterday, chances are they'll be there again today. If a tree had particularly nice fruit yesterday, chances are that fruit will be available today. Don't reinvent the wheel, and don't fix what isn't broken. Those are excellent laws to live by and poor laws to gamble by. Imagine picking a random spot to search for food every day or deciding to wear your warmest clothes for no good reason. Life would be exceptionally difficult. So difficult, in fact, that it's hard to check that programming at the door when entering a casino.

Confirmation Bias

The thing always happens that you really believe in; and the belief in a thing makes it happen.
Frank Lloyd Wright

Somewhere, someone once broke a mirror and experienced seven years of misfortune thereafter. Someone, somewhere died soon after crossing paths with a black cat. Flashes of lightning mean thunder is sure to follow. A particularly good hunt is followed by an evening of dancing around a bonfire. These are examples of *confirmation bias*, or the idea that A caused B to happen. Other examples include attempts to interpret observations and other information in a way that confirms one's preexisting beliefs, while discounting evidence to the contrary.

Chapter 5 explained the roots of confirmation bias as our innermost core beliefs. Confirmation bias is what keeps prey animals such as humans alive. Our core beliefs tell us what we need to know about the world and our places therein, and we follow those instructions faithfully, thanks to built-in emotional addictions. Beyond this, confirmation bias is an excellent learning tool. If I ate a certain plant yesterday and am sick today, then I probably shouldn't eat that plant in the future. On the contrary, if the plant was yummy and no ill effects occurred, then that plant is probably quite good to eat.

The drawbacks to confirmation bias lie mostly in the fact that modern life has evolved far too rapidly for our brains to keep up. Many of us have absolutely no reason to fear predation, and therefore no need for prey-based core belief systems. We can tell ourselves this rationally, but the simple truth is that we are emotional creatures beholden to million-year-old instincts. Far too many of us suffer from depression and/or baseless self doubts, thanks to confirmation bias. We just saw that casinos leverage this bias by displaying signs with the roulette wheel's last several results on the guess that players will use the previously discussed gambler's fallacy, which is simply a form of confirmation bias in action.

Seek and Ye Shall Find

It's a law of nature: A child with mountains of expensive toys will drop them all when presented with a cheap trinket or—even worse—a plain cardboard box. The initial burst of excitement soon fades, and the kid is once again open to receiving—if not actively begging for—something new. Attempts to reason with the child and point out the piles of great toys s/he already has are pointless.

If you would be a real seeker after truth, it is necessary that at least once in your life you doubt, as far as possible, all things.
Rene Descartes

Fast forward to adulthood when so many of us look eagerly forward to going out to dinner, a movie, our birthdays, getting a new motorcycle, etc. only to realize that life remains pretty much the same after attaining the object. I remember the eager anticipation I felt when buying my Yamaha FJ1200 motorcycle only to feel a tiny bit crestfallen as I drove it home. You've probably experienced similar things in your own life, where the joy of anticipation and searching for and looking forward to what you want is better than actually getting it.

Humans, like many animals, have a powerful mental drive to seek what we need to survive. A "seeking" circuit in our brains

> *Pleasure is a by-product of doing something that is worth doing. Therefore, do not seek pleasure as such. Pleasure comes of seeking something else, and comes by the way.*
> A. Lawrence Lowell

activates to both motivate us to begin the search and to keep us motivated until we satisfy our need. For example, a hungry person will keep looking until she or he has found food, at which point the seeking circuit shuts down. A hungry person persevering until locating food seems blatantly obvious. What isn't so blatantly obvious is the immense pleasure derived from the search itself.

Think about it: You're walking down a street and realize that you're hungry. Immediately, your mind fills with thoughts of the foods you love to eat and those available near your current location. You can almost taste the food and feel its warmth just by thinking about it. It's a happy feeling—and that happiness is what motivates you to keep on looking until you've found food. At that point, a different type of happiness may take over, such as the pleasant sense of well-being that comes with a full belly. A bit of a letdown may occur, which accounts for so-called buyer's remorse, where a person second-guesses a recent purchase.

Imagine if searching for food was an unpleasant experience. This may be the case for the billions of people trapped in poverty today, but keep in mind that poverty and social inequality are, like so much else in modern life, evolutionary novelties. All else being equal, the search for food is a very pleasant experience. If it wasn't, then people would probably put off looking as long as possible—too long, perhaps to have much chances at hunting, or even at finding edible plants. This could have serious consequences for survivability. The fact that searching for food (or any of life's needs) is so pleasurable helps ensure that we'll be out hunting and gathering long before we become too hungry to look. Put another way, the seeking instinct is nature's way of getting us off our rear ends before an emergency occurs, thus forestalling the emergency.

Quantity vs. Quality

> *Those who speak most of progress measure it by quantity and not by quality.*
> George Santayana

Be fruitful and multiply. That is nature's mandate, one most species ignore at their peril. Passing on one's genetic material to the next generation is one of the most fundamental drives there is for any species. No animal or plant will voluntarily curtail its reproduction when it has the ability and resources to increase its offspring. This is a very quantity-based approach, where all living beings are trying to reproduce as much as possible. Natural selection, competition for limited resources, and predation provide the quality side of the equation, ensuring—at least in theory—that only the best and brightest survive to reproduce in the future. Individual trait variations around a median value help ensure a level of genetic diversity that may help the species weather environmental changes. We've already explored this idea in detail throughout this book, and will continue to do so.

Technology has allowed humans to increase our food supply and boost our numbers far beyond what hunting and gathering alone could ever hope to support. We literally made the choice to keep increasing our food supply and our population, instead of limiting our needs by limiting our population. This stands to reason from an evolutionary perspective, because the species with the most offspring probably has the best chances for future survival. Furthermore, our awareness of Earth as a spherical planet with finite resources was not fully known until only a few hundred years ago, and the true limits of our planet's resources are only now starting to come to light. These are two more examples of events occurring far too rapidly for our brains to keep up.

Our unconscious choice to artificially increase our food supply and population is completely understandable, despite the tremendous associated costs such as social inequality, starvation, tyranny, organized violence (warfare), and a planet whose continued ability to support our species is coming ever further into question. I hope I'm not alone in wishing that the human species can find a way to care for the people alive today, while simultaneously taking steps to naturally reduce our population significantly. My hope is particularly fervent, because today's birthrates are highest in the areas that can least support new mouths to feed.

A much smaller population places far less strain on our planet's resources, helping ensure our planet's survival for future generations and the continuation of our imperative to pass on our genetic material. Reducing or even eliminating competition for resources can only improve everyone's quality of life, and reduce—or even eliminate—a large reason for going to war, an endeavor that has the potential to end all life on the only known habitable planet in the entire universe.

You are a product of your environment. So choose the environment that will best develop you toward your objective. Analyze your life in terms of its environment.
Clement Stone

Chapter 18

The Natural Savage at Home

Mid pleasures and palaces though we may roam, be it ever so humble, there's no place like home.
John Howard Payne

Shelter is both one of the six core life functions listed in Chapter 2 and highly dependent on safety, status, and food supplies. It is therefore appropriate to summarize what we've learned so far about predator avoidance, the need to fit into a group, and what and how we eat in the context of the average human's home life. The wide variety of human cultures on Earth seem to prefer an equally wide variety of types of shelters, from tents and grass huts to luxury high-rise condominiums. These seeming differences hide an amazing degree of sameness. Humans are, after all, only human.

Remember that humans are highly social creatures by design, because just about everything we do, from finding food to protecting ourselves, requires numbers. It should go without saying that virtually all human societies revolve around groups of adults. Free-ranging individuals tend to do so involuntarily (think *Castaway*, the Tom Hanks movie), while societies of children, such as depicted in *Lord Of The Flies* are pure—albeit plausible—fantasy. So, how do groups of adult humans live?

Home Sweet Home

Apes, including chimpanzees and bonobos, have no fixed home. They wander from place to place following available food, and building leafy nests in trees every night. Humans tend to settle down, even in nomadic societies, unless they are on a journey. What and how we eat offers some strong clues as to why this difference exists.

Remember that chimpanzees and bonobos are both arboreal and mostly vegetarian. An ape in a tree enjoys significant protection from cats and other land-based predators, and is too large for any bird to carry off (although we do still harbor the legacy of predation from the air, as described in Chapter 1). Fruits and plants can't move, meaning that hungry apes must travel in search of food. Once found, these fruits and plants (and occasional small meat such as insects or bush babies) can be eaten as-is without any need for preparation. Life for apes is one endless roving buffet whose pace is slow enough to allow mothers carrying infants to keep up easily.

Predators that bring down large animals have a different set of problems. Chases are fast, violent, and hazardous, no place for youngsters. These young must therefore be left behind, preferably in a den or other habitation that will help protect them from other predators. The young must also be fed, as must any caretakers, meaning that food needs to be transported home from the kill site. Territories may change and animals may move to follow the herds, but predators tend to have fixed bases of operation even if those bases aren't fixed for very long.

This is the true nature of home: It is the place of peace; the shelter, not only from injury, but from all terror, doubt and division.
John Ruskin

Humans fall into the predator camp as far as our living habits go. Unlike our ape cousins, we only rarely move on a daily basis, and exceptions usually occur because we're in the middle of a long journey. Whether it's a few days or a few years, we tend to remain in place for a fair amount of time. This gives hunters time to rest and recuperate between sorties, allows children to play under watchful eyes, and provides plenty of time for foraging. It also lets group members process food for eating and create/repair essential items, such as tools, weapons, and clothing. As an aside, this is why most exercise programs have some rest days built in—because the exercise/rest pattern is how we evolved. Further, as we saw in Chapter 16, our reasonably sedentary lifestyles combined with our omnivorous diet contributed to the advent of agriculture.

Designated Spaces

Architecture is the art of how to waste space.
John Howard Payne

I have either lived in, visited, or seen pictures of many different types of human habitations from tepees to huts, hotels, and houses, and lone dwellings to teeming cities. All of these types of shelter share a critical feature: Separate spaces for separate activities. Your own home contains a living area, sleeping area, eating area, garbage area, and excretion/bathing area. Each of these features may take wildly different forms, from walking off into the woods to do your business, to the communal toilets of the ancient Romans, and the multiple bathrooms per household in some modern homes.

Excrement is a strong reason for separating spaces by function. Feces from animals that eat meat smells much more strongly than feces from plant-eating animals, and also carries far more health risks to the animal itself. Apes don't, as a general rule, dispose of their own waste. A large percentage of them (over half, in some cases) even defecate in their nightly nests and sleep in it. Very few humans would voluntarily do such a thing without a really good reason such as being physically unable to move. Designating areas for defecating may have helped bring about agriculture, thanks to the large number of seeds deposited and fertilized in a small area.

Large quantities of food attract pests that must be kept at bay, if the food is to remain both plentiful and fit for human consumption. Concentrating food in designated areas also helps prevent contamination caused by rotting rood. For example, the Coast Miwok tribe of Northern California stored acorns wrapped in layers of hides in an elevated enclosure. Concentrating food and food processing in designated areas may also have helped bring about agriculture, thanks to the concentration of seeds in small areas.

My wife is always trying to get rid of me. The other day she told me to put the garbage out. I said to her I already did. She told me to go and keep an eye on it.
Rodney Dangerfield

Garbage heaps are relatively common archaeological finds. Discarding garbage in a designated area helps keep the living area clean and germ-free. It may also have helped agriculture, again because of the concentration of seeds in a relatively small area.

Agriculture allowed human populations to remain in the same place indefinitely, which compounded the sanitation problems faced by small bands. Aqueducts, sewers, granaries, dumps, zoning, and the like address these problems on the scale necessary to serve the entire population. Progress has not been linear, even in Western cultures. Ancient Roman cities featured clean water and efficient sewers that drained communal toilets. Centuries later, people in medieval Europe simply tossed their waste into the street.

Ants, termites, bees, and some colonial animals such as prairie dogs and meerkats also designate different spaces for different functions. Small prey animals that eat primarily plants and are too small to survive in the open benefit from forming large colonies that bear more than passing resemblance to human cities. If only human cities were as efficient! Termite mounds, for example, feature elaborate construction that maintains an internal temperature of about 87 degrees Fahrenheit and near 100% humidity (perfect for growing the fungus that the termites eat) without any machinery whatsoever.

Do not try to fight a lion if you are not one yourself.
African Proverb

Cavemen?

We often refer to our primitive ancestors as "cavemen," because of the archaeological evidence pointing to our having lived in caves. It is safe to assume that humans lived in caves. It is wrong to assume that we favored caves because other types of shelters such as huts, tents, and other structures built of mud, wood, hides, bone, etc. could have been built and later vanished without any traces. It is also highly improbable that caves alone could shelter the rising human population or that caves were the best choice in all situations.

Modern Savannas

Imagine the African savannas of our evolutionary youth, vast grasslands punctuated with trees and clumps of bushes. Now imagine a good percentage of the world's human living areas. The American Dream revolves around a house surrounded by a lawn and a few bushes and trees. No city is complete without parks that attempt to mimic the savanna by combining relatively small wooded areas with relatively large expanses of grass and open space. Few streets (open spaces) are complete without at least a few trees. Few homes are complete without at least a few plants. A home without plants, no matter how well appointed, looks cold and sterile—at least to me.

I speak for the trees, for the trees have no tongues.
Dr. Seuss

The Chinese art of *feng shui* (which literally translates as "wind-water") involves creating and locating spaces in order to create harmony between those spaces and their environments. Spiritual and *qi* (energy) considerations aside, wind and water are the two elements that can make life cold and miserable. Constructing shelters that take advantage of their location to provide maximum protection is a great idea.

Good Fences

The grass is not, in fact, always greener on the other side of the fence. Fences have nothing to do with it. The grass is greenest where it is watered. When crossing over fences, carry water with you and tend the grass wherever you may be.
Robert Fulghum

Humans and apes are territorial by nature, and hunting may have enhanced those instincts in humans, since areas with good hunting were in high demand. The tendency to stay put for at least a few days at a time probably also contributed to territoriality. Agriculture took the idea of *lebensraum* (literally "living room" or a place of one's own) to its logical extreme by providing a rationale for individual members of a society to own territory that required defending from both outsiders and insiders bent on helping themselves to ill-gotten goods. Over time, a series of complex rituals for peacefully entering another person's territory evolved. That bottle of wine you bring to dinner at a friend's house is more than just a friendly thing to do; it signifies that you know your place as a guest, that you come in peace, and that you'll leave when the time is right. Children are treated as extensions of their parents, and generally given roughly the same permissions and restrictions.

Modern humans are so territorial that we employ specialists (surveyors) to tell us exactly where properties begin and end, and build and maintain fences so that everyone knows where everyone else's boundaries lie. I know perfectly well that an ordinary garden fence is mostly ornamental since it won't stop an intruder. Still, I've either shied away from purchasing homes that lack fences or have erected them soon after the sale. My attitude is reflected in the old saying that "Good fences build good neighbors.."

Teeming Multitudes

Imagine the San Francisco Bay Area with only a few hundred inhabitants, a far cry from the roughly 7,000,000 that live there now. As I mentioned in Chapter 10, civilization occurred less than an eye blink ago in evolutionary terms. Despite this, humanity has, by most appearances, adapted rather well to crowded environments, where any individual might see more people in a single day than our ancient ancestors saw in a lifetime. Comprehensive sets of culture-specific maintenance keep the peace provided everyone has at least a modicum of available resources.

SIDS

Most mammal, monkey, and ape babies sleep with their parents, as do human babies around most of the world. Western cultures are different, because they advocate putting children (down to and including newborn infants) in bed alone. Is it any wonder that some children have problems sleeping? A lone baby is a prime target for predators in its own mind, because human brains have not adapted to our new, mostly predator-free reality. A baby asleep with its parents is a protected baby, which may make it less likely to be afraid of the dark and of monsters (predators). Interestingly enough, Sudden Infant Death Syndrome is far more common in societies that promote children sleeping alone. I have never heard of a child suffocating under or between its parents, and can only wonder whether fright is a risk factor for SIDS.

Having a baby's sweet face so close to your own, for so long a time as it takes to nurse 'em, is a great tonic for a sad soul.
Erica Eisdorfer

Sharing Our Homes

As we saw in Chapter 14, many of the interpersonal dynamics that exist in general society also exist within the home. Hierarchy, conflict, forgiveness, morals, sharing, cooperation, and scape-goating are just as common at home as they are out in the world. As we've seen throughout this book, humans walk a fine line between peace and conflict when dealing with one another. Our relationships have characteristics of both aggressive chimpanzees and peace-loving bonobos. One might think this combination would provide a mellowing effect; what it really does is broaden our options. The fundamental problem is that evolution wired each of us to strive to pass on our own genetic material to the next generation. Accomplishing this mission requires looking out for our own self interests above all else. The challenge is that sometimes our own interests are best served through restraint and cooperation. We therefore temper our conflicts with behavioral mechanisms that help

The only thing that will redeem mankind is cooperation.
Bertrand Russell

reduce the frequency and intensity of conflicts by preventing, recognizing, and righting real or alleged wrongs. These mechanisms have their roots in *morality*, which my dictionary defines as "a system of ideas of right and wrong conduct."

Morality

Men are more accountable for their motives, than for anything else; and primarily, morality consists in the motives, that is in the affections.
Archibald Alexander

The social contract we discussed in Chapter 10 stems from our need to band together for protection from predators. Evolution has built in powerful inhibitions against harming members of the group we depend on for our survival. Even the exceptions to that rule are far less likely to involve close blood relatives (remember that spouses aren't blood relatives). Put simply, being bad to another person feels bad. Guilt, sadness, shame, etc. are all emotional responses—and humans are emotional creatures first and foremost. We worry about people we've wronged, and may even volunteer for punishment. This has survival value, because it's hard to remain angry at someone who is volunteering for punishment. In fact, volunteering for punishment often results in either outright forgiveness or an actual punishment that is significantly less than the offender is willing to accept. This is a win-win because the aggrieved party obtains restitution while not having to feel guilty at having possibly caused excessive pain themselves.

Mixing one's own emotions with past experience allows one to extrapolate how one's actions will affect another. This is where one's sense of basic fairness comes from, which expands to form our concepts of right and wrong behaviors. This isn't all about helping our fellow man, though: Most humans are smart enough to know that treating someone unfairly is a great way to incite anger, which can become a grudge given enough provocation.

Aim above morality. Be not simply good, be good for something.
Henry David Thoreau

Just how far will we go to avoid arousing ill will in those around us? Monkeys in experimental settings will voluntarily starve themselves to avoid hurting another monkey and exhibit signs of distress at causing distress to others. As an aside, these experiments, in which a monkey that presses a lever to get food shocks another monkey, probably speak more to the cruelty of the experimenter and the artificial gap between humans and animals than to anything else. We don't like to cause pain because we know pain feels bad, and we don't want anyone to cause pain for us.

The same mental circuitry that allows morals also helps us survive. You may have seen herds of animals stampeding. One animal panicked, bolted, and the rest followed en masse, many of them probably unaware of what caused the initial alarm. It is notoriously easy to panic a crowd of animals, humans included. Plenty of people have died in fires and other disasters thanks to mass panic, this despite the rational knowledge that calm, orderly behavior would, in many cases, save all concerned. More proof that humans are anything but rational beings at heart.

As described in Chapter 10, humans normally have at least two sets of behavioral standards of fairness: One for fellow group members, and one for outsiders. Modern civilization with its maze of groups may require many such sets of standards for family, neighbors, friends, coworkers, strangers, foreigners, etc.

Cooperation Displays

Pay attention to people's body language the next time you head out to a place where people congregate, such as a park, restaurant, office, etc. Pay particular attention to the extent to which people in a given group unconsciously move to mimic or reflect each other's postures. As a general rule, people who mimic each other by reflecting "toward" each other are more comfortable around each other. Look for other gestures of openness, such as arms by one's sides or extended out to the side, open-legged stances or sitting positions, etc. These also signal comfort and trust. Closed-off gestures (crossed legs, folded arm, etc.) signal discomfort and/or distance. Two men sitting next to each other will often cross their legs away from each other, a not-so-subtle hint about the invisible line that few heterosexual men will cross with each other. Women will often sit with their legs tightly crossed and/or their hands guarding their genital regions. While you're at it, see how certain behaviors like yawns, checking the time, scratching an imagined itch, and more ripple across groups thanks to our built-in mimicry cells. And finally, listen to the conversations and see who is setting the tone for the rest of the group.

Whatever God's dream about man may be, it seems certain it cannot come true unless man cooperates.
Stella Terrill Mann

All of these behaviors give powerful clues about the level of cooperation, friendship, trust, rank, etc. each person has with everyone else in the group. Just about every person in every group you see knows her or his place and how they are being perceived by the others. Repeat this experiment at home, and you'll see that the same general rules apply. With a little practice, you'll be able to read a group's dynamics with amazing accuracy. This is no miracle. On the contrary, you're already doing it every time you interact with one or more people. The trick is to make yourself consciously aware of your existing subconscious knowledge.

Culture

Whenever I hear the word culture, I reach for my revolver.
Hanns Johst

Every household fits into a larger society that in turn embraces a *culture*, a matrix of values and behavioral standards adopted by the group. Culture can include language, cuisine, fashion, religion, taboos, fears, and more. Singer Janet Jackson briefly flashed one of her breasts on national television during Super Bowl XXXVIII in February of 2004, and the United States of America seemed to be apoplectic for days. Meanwhile, real and imagined violence is a staple of American life. Contrast that with the time I walked into a hotel room in Heidelburg, Germany, grabbed the remote, and clicked on the TV. Up came a talk show with footage of a group of naked skydivers. The camera zoomed in on one man, who was literally flapping in the breeze. The host and audience got in a hearty laugh, and the show moved on. Or the time I turned on the TV in Vancouver, British Columbia to see a skit performed by two men in penis costumes.

These contrasting attitudes to sex and violence are examples of cultural differences. Every culture has its unique mix. Culture promotes group unity and cooperation between group members. It also bolsters "us versus them" feelings and enhances group identity.

At face value, Earth hosts a dizzying assortment of cultures that are completely different from each other. Look under the hood, though, and you will see that these differences belie the fundamental sameness of humans everywhere. All cultures have norms that govern most waking activities as well as major life milestones such as birth, graduation, adulthood, marriage, parenthood, retirement, and death. This is usually a good thing. People who identify with and depend on one another normally get along better than those who don't.

224 | The Natural Savage
Discovering the Human Animal

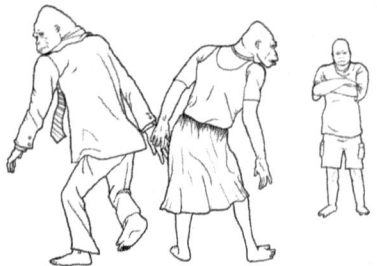

PART FIVE

Fruitful Multiplication

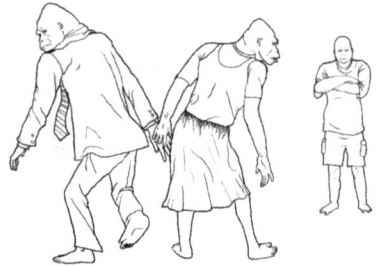

Chapter 19

Of Birds and Bees

> *I know nothing about sex because I was always married.*
> Zsa Zsa Gabor

If animals could talk, most any animal you asked would tell you that human sexuality is more than a little aberrant by their standards. Fully appreciating just how different human sexuality is requires setting aside our human-centric thinking to look at humans as just another animal among all animals. Remember the bell curve in Chapter 8? Only 3% of mammals are monogamous, and this percentage plummets when one considers the many millions of animal species in the world. This one trait alone places humans far beyond the +/-2 standard deviations from the mean that is generally considered to be the "normal" range for statistical purposes.

Comparing multiple sexual behaviors only increases the abnormality of human sexuality. Human females have evolved concealed ovulation, are sexually receptive at all times during their cycles, and experience *menopause* (cessation of fertility long before death). Most human copulation occurs horizontally, with the male on top of and facing the female (missionary position). We wear clothes, which cover up our built-in sexual signals, while at the same time wearing cosmetics, such as makeup and perfume, to enhance our sex appeal. Our sex normally takes place in private. We have recreational sex; in fact, we have most of our sex for the sheer fun of it, with no intention of reproducing. Modern technology has even allowed us to prevent conception from taking place. Any one of these is an oddity among animals, and no species besides *homo sapiens* possesses our truly unique package of sexual traits.

Why?

The best way to answer this question is to see how natural selection works, followed by a look at some of the many reproductive strategies animals use. We'll then look at some of the pressures facing human men and women and the lengths humans and animals will go to reproduce. This will provide the foundation we need to narrow our focus toward humans in the following chapters.

No Looking Back

Human sexuality has been evolving for millions of years, and most closely resembles a combination of chimpanzee and bonobo sexuality plus some additional, uniquely human traits. From how human eggs get fertilized to nursing infants and unequal parental investments, evolution has come too far down this particular path for it to change direction or backtrack, barring both some compelling reason to change and plenty of time to make that change. Put another way, I hope you like your evolved sexual traits, because those are the sexual traits you're stuck with.

Bisexuality immediately doubles your chances for a date on Saturday night.
Woody Allen

Why Sex?

All joking aside, sex is a costly, messy affair. It would be much easier to reproduce asexually like bacteria. Imagine being able to reproduce at will by simply splitting yourself in two. Mutations introduced by imperfect gene copying, and either encouraged or suppressed by environmental factors, drives evolution of the species. No more dating, courtship, sex, morning-after blues, child rearing, marriage, divorce, nothing. Asexual reproduction is simplicity itself. It makes sense for simple organisms that may not live long enough to have a reasonable chance of finding a suitable mate.

One of the aims of sexual union is procreation—the creation by reproduction of an image of itself, of the union.
Mortimer Adler

Complex organisms, such as humans and most animals, benefit from sexual reproduction, because it creates genetic diversity, particularly when long life spans are involved, because these factors limit opportunities for mutation and evolution.

Chapter 19
Of Birds and Bees

As we'll see in Chapter 21, humans tend to be strongly attracted to people whose genotypes are very different than their own, the more different the better. Mixing these disparate genotypes can give the child a stronger immune system, which helps the child pass on its genes and continue the parents' genetic legacy. At least that's how it works in theory. Environmental factors and changes can snuff out an otherwise promising genetic lineage, while creating conditions for a previously undesirable lineage to flourish. Those individuals best suited to current environmental conditions have more children. Remember that the purpose of life is to reproduce.

Sexual reproduction also creates conflict, because what's in each parent's genetic interests is often contradictory. The battle of the sexes is a very real phenomenon that exists for very good reasons, as we saw in Chapter 14. This conflict may seem awful, particularly in the heat of the moment, but it serves a strong evolutionary purpose: Those best able to mitigate or resolve their conflicts enjoy increased reproductive potential. It there comes as no surprise that humans couples who have devised mutually agreeable conflict-resolution strategies and/or who have similar ways of responding to various situations tend to remain together the longest. Even so, sexual reproduction in humans favors males who court many females, who respond by being highly selective. This explains (among many other things) why women placing personal ads seeking to meet men can usually count on receiving far more responses than men who place ads seeking women.

Men want sex. If men ruled the world, they could get sex anywhere, anytime. Restaurants would give you sex instead of breath mints on the way out. Gas stations would give sex with every fill-up. Banks would give sex to anyone who opened a checking account.
Scott Adams

Natural Selection

Life exists to beget life. Natural selection therefore favors adaptations that improve the odds of successful reproduction, defined here as both having more children and helping those children survive to have children of their own. These adaptations can be physical, such as breasts that mimic buttocks and thus serve as sexual signals. They can also be mental and/or psychological. For example, the average human would be hard pressed to recognize individual penguins in a huge Antarctic flock, let alone tell which chick belongs to which parent. You can bet your bottom dollar that penguins can recognize each other and discern their young in a penguin crowd, just like humans can spot a familiar face in a human crowd. I can't imagine any species whose children require parental care lacking the ability to recognize individuals, because that would lead to parents raising just any child, a clear no-no in evolutionary terms under most circumstances.

This is a quantity game; parents who can have the most offspring are theoretically the ones who are best suited to their current environment. Any change in the environment could strip a given set of genes of its favored status and replace it with a different set. Extinction occurs when insufficient numbers of individuals with any sets of genetic traits exist to repopulate the species. How exactly parents manage to have the most possible children is wide open to question—and therein lie the seeds of parental conflict. Please review Chapter 14 for more details.

Natural selection chooses among currently available options. Two parents represent two sets of genetic traits, or options, and their child represents a third trait that combines the first two. This is the menu open to natural selection. Two parents won't create a child with completely different genetic material, meaning that natural selection cannot simply invent a new option, any more than patrons at most restaurants can cook their own food or invent special dishes.

I must take a short detour to explain that words such as *choose* and *invent* usually imply deliberate, conscious action. Applied to natural selection, these words can imply the presence of an unseen super-intellect, which can be referred to as a god or higher power. I don't necessarily mean to imply this, because no conscious action, blueprint, or designer may be necessary for natural selection to occur in at least some—if not most or even all—cases. For example, the mix of parental genes inside a child could predispose it to store extra body fat. If the environment cools, then this child and anyone else with a similar trait may be better able to survive and have children of their own. If the climate warms, then this child and others like it may be unable to survive long enough to reproduce. This "choice" may be an example of natural selection with no guiding intellect or design behind it. Then again, it could be proof positive that some super-intellect (God) is indeed real. As for "invention," imperfections in gene copying can introduce *mutations*, or changes in the child's genes that are not present in

No matter what a woman looks like, if she's confident, she's sexy.
Paris Hilton

Nature is an infinite sphere of which the center is everywhere and the circumference nowhere.
Blaise Pascal

Chapter 19
Of Birds and Bees

> *All women's issues are to some degree men's issues and all men's issues are to some degree women's issues because when either sex wins unilaterally both sexes lose.*
> Warren Farrell

the parents' genes. Imagine transcribing a letter and omitting or misspelling a word. That is completely different than writing a completely new letter from scratch without trying to faithfully copy the original.

I'll briefly discuss evolution, intelligent design, and creation in Chapter 26. Another book of mine in this series, *The Divine Savage*, is entirely devoted to this discussion.

Reproductive Strategies

Let's take a look at some of the many ways animals reproduce.

No Contact

> *Women have more to offer this world than just a fallopian tube. Nothing is going to change until you quit looking at us as just sperm receptacles.*
> Barbara Hall

Many female fish lay eggs directly into the water, whereupon the male releases a cloud of sperm. There is no actual contact between the mates. This works well, because many of the shapes best suited for various underwater environments may not lend themselves to actual coitus. Some fish species guard their eggs and protect the hatched young, while others swim away, leaving the eggs to their fate. It makes little sense to guard a large clutch of eggs where the sheer numbers all but guarantee successful reproduction. On the other hand, fish who lay fewer eggs and/or whose eggs have longer gestation periods may gain by hanging around to protect their offspring. The basic formula is simple: Investing energy in any one egg that results in less protection for other eggs and lower resulting odds of reproduction makes no sense. Remember, evolution is a numbers game.

Single Queen

> *She moves a goddess, and she looks a queen.*
> The Iliad

Ants, bees, and termites have a unique caste system, where sterile workers tend to the young, grow food (ants and termites were the world's first farmers), and keep the colony clean and maintained. Soldiers, also sterile, guard the colony and invade other colonies. Fertile males mate with a queen, who then establishes a new colony, settles down (she will never move again after this), and starts cranking out eggs. Eliminating reproductive issues from most of the population provides an absolutely loyal and efficient workforce that in turn helps ensure the great number of births and survivors, a few of whom will go on to become fertile males or queens. Eggs and queens are carefully guarded, and the young are fed and raised.

Polygyny

Animals who give birth to live offspring may benefit from *polygyny*, where one male maintains a harem of several females. This benefits the females, since they don't have to worry about rival male advances, giving them more time to look after their young. Their reproductive ability is not slowed down by sharing the male, since they may only give birth to one child at a time (such as walruses). Dominant males win more mates, and thus an increased opportunity to pass on their genetic material. Battles for supremacy tend to go to the largest males, so the pressure is on males to grow as large as possible. Male walruses can weigh twice as much as females, and possibly even more. The average size of a harem depends on the ratio of male/female size.

Human males are larger and approximately 20% heavier than human females. Humans are therefore at least mildly polygynous by nature. We'll return to this topic in Chapter 22.

All marriages are mixed marriages.
Chantal Saperstein

Serial Monogamy

Most birds form pair bonds for a single season, although some do mate for life. Eggs and chicks are extremely vulnerable to both environmental factors like temperature and predators. One parent must therefore guard the nest at all times, while the other finds food for the brood. Some species have both parents taking turns on the nest, while other birds have well-designated gender roles. Chicks generally become mature enough to no longer need constant watching after one season, so it makes little sense to limit one's reproductive potential by sticking around too long.

It may be to one parent's advantage to pass the child-rearing buck to the other parent and go off to reproduce some more. The parent with the most invested in the outcome (typically the female) has a harder time shirking its duty. This may leave males free to leave the nest ostensibly on a hunting run, and simply mate with another female, who is then committed to

The trouble with my wife is that she is a whore in the kitchen and a cook in the bed.
Geoffrey Gorer

raising the chick. Birds can't abort their eggs, unlike human fetuses.

Monogamy

> *A simple enough pleasure, surely, to have breakfast alone with one's husband, but how seldom married people in the midst of life achieve it.*
> Anne Morrow Lindbergh

Parents who both invest heavily their children may benefit from monogamy. Human babies are born helpless, and remain more or less helpless for a long time—over two decades, once one factors in a potential college education. A human child can spend between 1/4 and 1/3 of its life preparing for adulthood—a fraction that may be relatively constant throughout our evolutionary history as life spans increased along with the complexity of education. Children with only one parent face a significantly higher risk of starvation and predation, particularly when that parent is a mother who is not as well suited for fighting off enemies as the father. Thus, if the father wants to see his children grow to reproduce and bear grandchildren, he'd better be by the mother's side doing his part.

It is no coincidence that human sexual signals and behaviors evolved to fit this exact scenario. Why do women conceal ovulation, and why are they sexually receptive throughout their entire fertility cycles (unlike chimpanzees and bonobos who do neither)? Because sex is fun. It has to be, in order to motivate us to try and reproduce as much as possible. A person's body responds in exactly the same manner whether the sex is procreationally or recreationally intended. The male gets to enjoy sex all the time, and thus has little need to seek satisfaction elsewhere. The female get to enjoy sex all the time, while protecting her enormous investment that begins with eggs that are roughly 1,000,000 times the size of sperm, and continues through a risky pregnancy (see Chapter 8), nursing, and carrying the baby.

Polyandry

> *When a woman marries again, it is because she detested her first husband. When a man marries again, it is because he adored his first wife. Women try their luck; men risk theirs.*
> Oscar Wilde

One female with more than one male (as seen in the movie *Paint Your Wagon*) conveys significant advantages for the female, because she enjoys increased protection and a more reliable food supply. It does nothing, however, to help her reproduce more rapidly or nurse more than 1-2 infants at once. All of the males who did not sire the child lose the great game of life (passing on the almighty DNA), because they are helping another male's genes succeed at their own expense. Which brings me to...

Cuckoldry

Any animal, humans included, that can sire a child and then get another individual to raise it as its own has won the evolutionary lottery, because it gets to pass on its genetic material without doing any of the work. The cuckold (animal raising the other child as its own) is at a double loss, because it is both not passing on its own genes and working as if it was. Cuckoldry occurs far more often than one might think, even among humans.

Dancing begets warmth, which is the parent of wantonness. It is, Sir, the great grandfather of cuckoldry.
Henry Fielding

Don't confuse cuckoldry (raising someone else's genes thinking they're your own) and *adoption* (deliberately raising someone else's genes, which also occurs surprisingly often, including with me). Both of these are also distinct from relatives helping in child care, because by doing so, they are helping pass on at least some of their own genes.

Incest

I've already mentioned that humans tend to be most strongly attracted to people whose genotypes are the opposite of their own. Does your significant other have a unique scent that makes you weak at the knees? You're smelling her or his opposite genotype, and your body is telling you that this mating has a strong likelihood of producing the kind of robust offspring that would be well able to reproduce in its own turn.

A sympathetic Scot summed it all up very neatly in the remark, "You should make a point of trying every experience once, excepting incest and folk dancing."
Sir Arnold Bax

If partnering with someone of the opposite genotype is good, then it stands to reason that mating with someone of a similar genotype is bad. The people with the closest genotypes are our blood relatives: siblings, parents, uncles and aunts, cousins, etc. *Incest* (sexual relations with blood relatives) carries a heightened risk of birth defects, mental retardation, etc. Small wonder that most (if not all) animals, humans included, have a built-in taboo against committing incest.

Our built-in incest taboo is so strong, that it can even extend to non-relatives who were raised near us as children. A survey of over 2,700 Israeli marriages turned up only 13 (just under half a percent) from children raised in the same kibbutz—and all of these children had moved in after age 6. A prepubescent human is not ready for sex, but she or he already knows who not to mate with.

Internal vs. External Fertilization

> *Probable-Possible, my black hen, she lays eggs in the Relative When. She doesn't lay eggs in the Positive Now because she's unable to postulate how.*
> Frederick Windsor

Fish that simply release eggs and sperm into the water (as well as other animals that fertilize their eggs externally) have invested roughly equally in the reproduction process. Internal fertilization requires a large up-front investment on the mother's part, in the form of specialized structures (genitalia) that allow mating and insemination to occur. Animals that give live birth (as opposed to birds that lay fertilized eggs) require extensive specialized organs, including genitalia, uterus, and placenta.

External fertilization gives both parents an equal opportunity to either ditch the other to watch the offspring, or even to be cuckolded. Female animals that either lay fertilized eggs or give birth to live offspring can be absolutely certain that they are, in fact, the mother, with rare exception. Ensuring paternal certainty is another consideration altogether, one that in humans has forced almost all societies to adopt complex laws designed to ensure marital fidelity and even premarital virginity, particularly of females.

Mixed Strategies

> *The most dangerous strategy is to jump a chasm in two leaps.*
> Benjamin Disraeli

Internal fertilization sets up the inherently unequal investments described in Chapter 14, making males more likely to have to pitch in, because it's almost impossible for the female to raise the young solo. Humans are an extreme example, because our children remain helpless after weaning, but even that isn't enough to completely level the playing field, nor to completely align the parents' interests (as evidenced by human males being larger than women). The man, like some male birds, will be tempted to let his eye—and more—wander somewhat. Polygyny and deception can work, as we'll see in Chapter 21, because the fraud may not be discovered in time. This "mostly monogamous with occasional side trips" approach is called a *mixed reproductive strategy*. It can work, but it's a pretty risky gamble.

I hate to say this, ladies, but this fundamental male drive won't disappear any time soon. It would be wonderful if all fathers stuck by their mates and devoted themselves wholeheartedly to one family and one set of children. The problem is that natural variances in temperament (see Chapter 8) would produce men who were more likely to cheat. These men would have more offspring, who would eventually crowd out their faithful brethren. Evolution is a numbers game. Period.

Sexual Cannibalism

If you're a man, imagine having sex with a woman. As you finish, you experience an uncontrollable urge to lean down and offer yourself to your mate, who proceeds to kill and eat you, perhaps while you're still ejaculating. If you're a woman, imagine feeling uncontrollably hungry, so hungry that you start eating your mate while still having sex. It's OK to be freaked out and a little disgusted at this visual. After all, you're human. This scenario makes no sense to you, because your life span is measured in decades, giving you plenty of time to find food. Besides, killing the male during sex makes it awfully hard for the mother to raise her child, and reduces the child's chances of survival.

Now put yourself in the position of an insect (such as some species of mantis). Your life span is measured in days. After sex, the mother lays a bunch of eggs and flies off to die, leaving the eggs to fend for themselves. There is no need for parental involvement beyond copulation and egg laying. The short life span makes potential mates a rare commodity, meaning you'd best do it with the first candidate who happens along. Producing and laying eggs consumes a lot of energy, particularly when your reproductive strategy involves laying as many eggs as possible in the ultimate numbers-game hope that a precious few will survive.

Life exists for the purpose of begetting the next generation. If you're the father, your entire reason for being ends the moment you finish mating. Remaining alive would be a waste of resources, and nature wastes nothing. If you're the mother, then you need a big meal to gather the energy you'll need to put into growing all those eggs. Flying around trying to find food consumes precious energy in the same way that every failed hunt increase a predator's risk of starvation. You need a big meal, and the big male you just got done mating with has

Cannibals prefer those who have no spines.
Stanislaw Lem

I am just another insect on the windshield of life.
Don Mashak

no more reason to live. Seen in this light, not eating your mate becomes the foolish choice.

If you think this sounds cruel, consider that the only reason you're alive to read this book is because plants and animals died to give you nutrition. All life requires death in order to continue, from organic compounds released by decomposing corpses that feed plants, to the antelope that feeds a tiger. Nature does not care for any individual in the slightest; it cares for the species as a whole, meaning that it cares for the genes. You, dear reader, are little more than a mobile laboratory built for the sole purpose of mixing and remixing DNA, and ensuring that your unique mixture lives to mix again. Everything else—including reading this book—is fluff, as far as dear old Mother Nature is concerned (at least on the purely biological level that this particular book is concerned with).

This holds true even if you never actually have children. The best moments of your life will happen when you are with someone in a pair-bonded relationship as I am with Jennifer. And just what are you doing in said relationship? Why, you're going through the reproductive motions, of course!

Productivity

If we are going to teach creation science as an alternative to evolution, then we should also teach the stork theory as an alternative to biological reproduction.
Judith Hayes

The average human woman can produce somewhere between 10 (breast-fed) and 30 (bottle-fed) children in a lifetime despite having about 400 available eggs, because she can't incubate or breast-feed more than one uterus load of 1-2 (usually) children at once. Meanwhile, the man is ready at a moment's notice, and can theoretically sire thousands of children. Staying around to help raise the kids puts a serious crimp in his style, as I just noted above. The exception is when he can be positive that the children he raising are his genetic legacy. Cuckoldry is a real risk for human fathers. Human mothers have zero risk of cuckoldry. If the baby came out of a woman's body, it's hers, period. The joy of this certainty may at times be tempered by her utter inability to hand off caring for her children.

Vasopressin

Prairie voles (*microtus ochrogaster*) are small rodents that resemble mice with small ears and rotund bodies. They are roughly the size of small to medium rats. What makes this particular species of vole noteworthy is its human-like mating habits, right down to the nominal monogamy (mostly monogamous with occasional side action). Males guard their mates and young, fetch food, clean the burrow, and more. They do this thanks to *vasopressin*, a hormone that increases both aggression and fathering instincts. The more vasopressin, the more faithful the male vole is likely to be.

Humans also have vasopressin. Early experiments seem to indicate that vasopressin in human males may perform the same functions. A surge of vasopressin during sex seems to enhance pair-bonding, while also triggering aggression towards other males. This jealousy causes the male to attempt to keep other males away from his woman while being very devoted to her. I can't help wondering whether any relationship exists between excessive vasopressin and domestic abuse.

Rolling the Dice

Your head may still be reeling from what you just read about sexual cannibalism being just one more example of the reproduction imperative. Sexual cannibalism may be relatively rare, but consider that starving animals will routinely choose sex over food when given the choice. Deliberately pushing an animal to the brink of starvation and then giving it a choice between sex and food strikes me as a particularly awful experiment, but it does prove my point about the absolutely central role that reproduction plays in all of our lives.

> *The pursuit of truth and beauty is a sphere of activity in which we are permitted to remain children all our lives.*
> Albert Einstein

Let's look at sex in the context of all we've discussed so far. Sex is expensive—and I don't just mean fancy dinners and jewelry! It consumes a lot of energy. It consumes a lot of time that could go toward finding food. Humans normally mate lying down, making them easy targets for roaming predators. Sex can even trigger heart attacks and other health issues. Fighting over sex and mates can be injurious or even fatal. Sex isn't cheap. Then again, human sex is a lot cheaper than, say, mantis sex.

Chapter 20

Sexy Monkeys

It is disturbing to discover in oneself these curious revelations of the validity of the Darwinian theory. If it is true that we have sprung from the ape, there are occasions when my own spring appears not to have been very far.
Cornelia Otis Skinner

Humans and bonobos both have valid claims to the title of "sexiest ape alive." Most humans don't hold a candle to the bonobo when it comes to sexual frequency and variation. For example, bonobos will readily engage in homosexual activity, something a relatively small percentage of humans do at all, much less do with any regularity. On the other hand, the human penis is about twice as big as a chimpanzee or bonobo penis, and almost four times larger than the average gorilla penis. Human females, unlike bonobo females, remain receptive throughout their menstrual cycles, and even during pregnancy. This, coupled with the human male's small testicles (compared to chimpanzee and bonobo), seems to indicate that humans are biologically designed to engage in recreational sex, that is, sexual activity with no intention of reproduction.

By contrast, bonobos have so much sex with so many different partners, that the male with the largest testicles has the best chance of passing on his all-important genes, because he can produce more sperm than his competitors. This feature is not part of human anatomy, and yet females are constantly displaying sexual signals (breasts, lips, buttocks, etc.). Clothing may cover some of the actual body parts, but both their shape and allure remain.

If sex is designed to pass on genetic material, then why are humans seemingly built for sex that does not directly achieve this end, particularly when sex is one of the costliest activities there is, in terms of calories and increased predation risk? Why do human females display sexual signals regardless of fertility, and why are men consistently attracted to them? Why would evolution go out of its way to make sex as enjoyable as humanly possible (no pun), so much so that we routinely engage in the recreational sex our bodies are apparently built to have? And why does this recreational sex tend to be better (much better, in fact) with partners who share an emotional bond?

Fidelity

The short answer is that human sex, especially sex between emotionally bonded partners, is fun (and women are sexually receptive at all times, even during pregnancy) because drifting from partner to partner is just not an option for our species. The longer answer is that the neoteny we discussed in Chapter 8 results in infants and children who are utterly dependent on their parents. Even today, children with two active parents have a much better chance of success—a fact not lost on family courts that routinely order one parent to make child support payments. How do you keep two animals together for years? Make them fall in love, and let them enjoy sex at least as much as animals who mate solely for reproductive reasons. Seen in this light, so-called recreational sex takes on a decidedly reproductive function.

The advent of contraception and sex with no intention of procreation takes full advantage of this biological reality, to the point where many people have several sexual partners before "settling down" to marry and have children. Attitudes towards premarital sex differ, from those who advocate sowing as many wild oats as possible (including sex with prospective spouses), to remaining chaste until marriage. Those who favor the former have the attitude that "it's a good idea to taste the milk before buying the cow" (to quote a longtime friend of mine), while those with more conservative attitudes figure that the couple will have no basis for comparison, and therefore no reason to complain about the quality of their marital liaisons.

What a fuss people make about fidelity! Why, even in love it is purely a question for physiology. It has nothing to do with our own will. Young men want to be faithful, and are not; old men want to be faithless, and cannot. That is all one can say.

Oscar Wilde

> *Another of our highly prized virtues is fidelity. We are immensely pleased with ourselves when we are faithful.*
> Ida R. Wylie

> *Nearly all marriages, even happy ones, are mistakes: in the sense that almost certainly (in a more perfect world, or even with a little more care in this very imperfect one) both partners might be found more suitable mates. But the real soul-mate is the one you are actually married to.*
> J.R.R. Tolkien

As someone who believes in evolution, I must set aside my spiritual and moral ideas long enough to point out that this lifetime may be the only one we get, leaving nothing to be gained by a life of chastity.

Back to the main topic, biology designed us to keep our relationships, at least for the most part. Remember the uneven reproductive investments we discussed in Chapter 14? Men are far more interested in women sexually, and are far more likely to have casual sex because each such "fire and forget" encounter is a chance at passing on the genes and having someone else do the actual raising. Why raise your own children if you can find a cuckold to do it for you? This explains why men are far more likely to have extramarital affairs that "don't mean anything." It is perfectly possible for a man to have an extramarital affair without losing neither love for his wife, nor devotion to his children. Women have extramarital sex for much different reasons. Their affairs usually stem from the desire to find a better provider for themselves, and possibly for any children they may have. If their current mate is failing as a provider, then replacing him may enhance the woman's chances to have a child and/or successfully rear any current children. While the man's primary contribution is economic in most cases, women also require (and fully deserve) emotional support and help around the home. Put another way, the man who does as much as possible to even out the couple's inherently uneven reproductive investments stands to remain favored. So, guys, the next time she tells you to take out the trash...

Various estimates place the chances of first marriages ending in divorce at between roughly 40 and 50 percent, with higher percentages for second and subsequent marriages. On average, marriages last between eight and 10 years. Ten years old is roughly when boys and girls begin the biological transformation from child to adult known as *puberty*. Human life expectancy fluctuated between 18 and 20 years between the Neanderthal and Bronze Age periods, only surpassing 30 years with the onset of civilization. Remember that we have not yet evolved to deal with civilization; it therefore makes sense that humans would rear their young for 10 years, after which the children would leave home to begin raising their own children. Couples gain no evolutionary advantages by remaining together longer than necessary to raise children to the point where they can reproduce on their own. These numbers don't take modern life expectancy (a by-product of civilization) into account, nor do they account for such modern considerations as sending children off to college or laws regarding the age at which a person may consent to have sex.

The conclusion seems obvious: Human sexuality evolved to help create and maintain a (mostly) monogamous pair bond that endures long enough to rear children to biological maturity, after which there is little reason for the parents to remain together. The religiously-inspired "til death do us part" ideal seems completely at odds with biology. I have to wonder about the people who obey this directive. How many of them remain married from a sense of duty and rightness? How many of them, with their last dying breaths, reflect upon their marriages and feel—truly feel—that they have done the right thing? I am not denouncing the institution of marriage; I am simply wondering how many of the expectations we take for granted have their basis in evolution.

God created sex. Priests created marriage.
Voltaire

Attraction

So what exactly attracts people to each other? This simple question has many complex answers. I break attraction down into physical (how someone looks), biochemical (genotypes and pheromones), demographics/psychographics (ethnicity, beliefs, etc.), handicaps (how someone overcomes real or contrived obstacles), and economics (resources at one's disposal).

I know love and lust don't always keep the same company.
Stephenie Meyer

Physical

Get this: One of my oldest friends has a thriving career as a *doula*, an assistant to new mothers. Part of her training includes teaching infants how to latch on to their mothers' breasts so they can draw milk. I was more than a little incredulous when she first told me about this. After all, haven't children been nursing for millions of years? Turns out that connecting to a human breast for feeding purposes is a little harder than one might think ,and some babies do actually have serious problems getting lunch. The problem is that human breasts are designed to mimic buttocks, the cleft between them bearing more than slight resemblance to the genital region. Chimpanzee and bonobos walk on all fours and exhibit prominent genital swellings when fertile, so they have no need of large rounded breasts. Walking upright renders a woman's genitals all but invisible. Breasts adapted to mimic the buttocks/genitals at the expense of making nursing somewhat more challenging. This may help explain why breasts fascinate so many men. As alluring as breasts are, though, they alone are no match for a host of other physical traits, including:

- *Lips.* Soft folds of pink everted flesh evoke the vagina, which reddens when aroused. Cosmetics manufacturers exploit this by selling lipstick in every imaginable shade of red.

- *Healthy glow.* Ruddy cheeks can be a sign of health, recent exercise, or sexual arousal.

- *Smooth, evenly-colored skin.* Diseases from acne and chickenpox to skin cancer, allergies, and a host of other ailments and injuries leave their marks on the skin, which can become discolored (jaundiced, ashen, etc.), unevenly colored, mottled, scarred, bruised, etc. Smooth, even skin is a sign of health and is therefore attractive. This is important because humans reproduce very slowly compared to other species. We rely on sex to mix our genes and keep germs guessing.

- *Symmetry of features.* Do both sides of a face mirror the other? Asymmetry can indicate disease or deformity.

- *Youth.* The older a woman gets, the less likely her children are to survive. A woman in her twenties has about a 1/2,000 (0.05%) chance of having a child with Down's syndrome. This increases to 10% for a woman in her late forties. Is it any wonder that a man's second wife is usually younger than her first? Besides, a younger mate has a longer time in which to possibly bear more children than an older person.

- *Age.* Women tend to be attracted to men who are anywhere from slightly to significantly older than they are. Older men tend to hold higher status and command more resources than younger men who often work for or otherwise defer to their elders. This may help explain

---

Nay, but Jack, such eyes! such eyes! so innocently wild! so bashfully irresolute! Not a glance but speaks and kindles some thought of love! Then, Jack, her cheeks! her cheeks, Jack! so deeply blushing at the insinuations of her tell-tale eyes! Then, Jack, her lips! O, Jack, lips smiling at their own discretion! and, if not smiling, more sweetly pouting —more lovely in sullenness! Then, Jack, her neck! O, Jack, Jack!
Richard Brinsley Sheridan

why a middle-aged man with a full head of hair can be such hot property.

- *Physical fitness.* Let's face it, someone who is at least reasonably fit has a far better chance of living longer than someone who is absolutely sedentary and/or grossly overweight.

- *Finger length.* More people care about length of middle finger than hair/eye color, etc.

- *Waist-hip ratio.* I've read a lot about the supposedly ideal 0.7 waist/hip ratio (where a 26-inch waist would correspond with 37-inch hips, awfully close to the "ideal" 36-26-36 we hear so much about). Hips are certainly attractive. A woman with wide hips will presumably have an easier time during childbirth and may even be able to reproduce a bit faster than her narrow peers because of the potential for reduced birth-caused trauma. This convincing argument fails to account for cultural differences. For example, a thin woman in a land of plenty may demonstrate restraint and good judgment while a large woman in an area lacking resources may indicate a strong survival instinct.

- *Parental measurements.* Believe it or not, the relative length of a potential mate's middle finger to their other fingers may play a bigger role than some or all of the traits we just discussed. People notice such subtle details without being aware of noticing. We tend to be attracted to people whose finger-length ratios are the same as our parents. On a different level, my own body bears more than a passing resemblance to my first wife's father.

> *Hands promiscuously applied, round the slight waist, or down the glowing side.*
> Lord Byron

Biochemical

Remember the incest taboo I mentioned in Chapter 19 that is so strong that children raised in the same kibbutz almost never marry each other, despite not being related? This may have everything to do with smell. No study that I'm aware of has proven the existence of human *pheromones* (chemical cues designed to trigger various behaviors in others, including attraction or rejection). Still, there is ample evidence that smell plays an extremely significant role in attraction. Someone raised around a given smell since birth (or very young youth)

> *Smell is a potent wizard that transports you across thousand of miles and all the years you have lived.*
> Helen Keller

may come to associate that smell with "family" and hence with incest, which is something to avoid. Why? Ah, now this is where it gets interesting.

Every single human is radiating an enormous amount of chemicals such as sweat, dead skin cells, hairs, and oils. Each of these items has a smell that is unique, both to the type of item (sweat has a different smell than oil) and to the person secreting that item. I once watched a show about a fugitive who was tracked down many miles of interstate and then off an exit and into the woods—by a bloodhound who was smelling the skin cells shed by the fugitive and blown out the car's vents! Every single person traveling that interstate was venting her or his skin cells out their own car windows, yet the bloodhound tracked the fugitive's scent without distraction. Bottom line: humans smell.

Here am I: at one stroke incestuous, adulteress, sodomite, and all that in a girl who only lost her maidenhead today! What progress, my friends... with what rapidity I advance along the thorny road of vice!
Marquis De Sade

Incest is a bad idea, because our relatives have extremely similar genes, the mix of which confers few benefits and many risks. The average modern human is hardly a trained geneticist, and the field itself didn't even exist a century ago. Still, we've managed to do a great job of not having kids with our close relatives, for the most part. It seems that the smells we give off carry information about our *genotype*, or individual genetic makeup. If incest is bad because the genes are too similar, then it follows that someone whose genotype is as different as possible would make an ideal match. This is precisely what happens. My partner Jennifer loves smelling my (bald) head. I can only surmise that she and I have very(!) different genotypes. Why is this good? Because mixing radically different genotypes from two healthy people carries little risk of inducing defects, and provides a great platform for creating a strong immune system in any offspring. A child with a strong immune system has a much better chance of living long enough to... wait for it... pass on the parents' genes to the next generation.

But back to the kibbutz. Surely some of the many thousands of children there must have very different genotypes. So why were there practically no marriages? Again, humans aren't geneticists. The answer could be as simple as an instinctual switch that says "smell since birth = family = incest = bad." If everyone in your hunter-gatherer tribe smells like family, that could encourage you to find a different tribe, and may explain why intermarriage between tribes remains such a sought-after event. Our closest evolutionary cousins, the chimpanzees and bonobos, migrate away from their birth groups. Coincidence? I think not. The old adage of "opposites attract" is spot-on in many ways.

Hormones also come into play. Falling in love, having sex, and even being around one's partner alters levels of *dopamine* and *oxytocin*, to name just two. In very rough terms, dopamine triggers so-called pleasure centers in our brains that help motivate us to seek rewards. Oxytocin is related to sexual arousal, and plays a role in pair-bonding and monogamy. It is chemically similar to vasopressin, which I mentioned in Chapter 19.

Demographics/Psychographics

The human tendency to remain together and co-parent children long after copulation is another rarity among animals. Physical and chemical attraction can trigger lust, sex, and an initial falling-in-love period, but that is not enough to sustain a pair bond (long-term relationship/marriage) over the long haul. *Demographics* (traits such as income, ethnicity, age, location, etc.) and *psychographics* (beliefs, attitudes, prejudices, opinions, etc.) are tremendously important. Most couples share the same religion, ethnicity, race, class, age, and political views. Personality and intelligence are also key components that can determine attraction (or not). Guess what: the relative importance of such physical traits as ratio of middle finger length falls between religion/ethnicity/race/etc. and personality/intelligence in importance as a factor in attraction!

Love is a force more formidable than any other. It is invisible, it cannot be seen or measured, yet it is powerful enough to transform you in a moment, and offer you more joy than any material possession could.
Barbara De Angelis

Ask most anyone what they seek in an ideal mate, and they'll probably rank "sense of humor" at or near the top of the list. Is this true? Well, in a sense, yes. Sort of. All else being equal, two people with compatible senses of humor are far more likely to remain together over the long haul. Humor has the capacity to amuse, enlighten, annoy, offend, and cause emotional pain. It also provides a window into the jokester's innermost thoughts, feelings, prejudices, and beliefs. Learning when different humor is appropriate is a critical social skill to master. Two people with compatible sense of humor therefore share compatible psychographics.

Chapter 20
Sexy Monkeys

In Chapters 4 and 5, we saw how people only see schemas of reality because of their core beliefs filtering raw perceptual data. Humans are remarkably consistent creatures, meaning we will tend to respond to the same stimulus in roughly the same way every time. Some psychologists refer to this pattern of identical reactions as a *script*, with everyone following their own internal scripts for every situation they encounter. It follows that people will probably be most strongly attracted to those who have similar scripts. Someone who freaks out at the slightest provocation is likely to annoy someone with a cooler head. On the other hand, dealing with someone else's drama can be a great way to avoid dealing with one's own inner demons. I've been guilty of that myself from time to time. Another phenomenon that sometimes happens is called "marrying one's parents," where a person will find a partner whose personality is either close to—or diametrically opposed to—some or all of the person's parents. In general, however, the happiest couples seem to have similar attitudes towards life, as well as similar ways of coping with/resolving problems and long-term goals. This upholds the common observation that "like marries like."

Demographics and psychographics are large factors in attraction, particularly long-term attraction. They don't, however, seem to play too big a role during the initial attraction/lust/falling-in-love phase of a relationship. Several longtime friends of mine have subscribed to an online dating service that matches people based on multiple "personality dimensions," and all of them report the same phenomenon: They get along fabulously with most of the people they meet, and find their matches to be extremely compatible personality-wise. My friends uniformly report that they could see themselves being great friends with many of the people they meet through this service. The catch is that they don't often feel that initial "spark," which is primarily triggered by physical and biochemical factors. I've seriously toyed with the idea of launching an online dating service that would analyze members' finger measurements and genotype, and add those components to prospective matches. I suspect that any given member would receive far fewer potential matches compared to my competitors, but that the percentage of meetings resulting in strong attraction would be markedly higher.

Fantasy love is much better than reality love. Never doing it is very exciting. The most exciting attractions are between two opposites that never meet.
Andy Warhol

Given what we've just discussed, it should be readily apparent that the separate ideas of "opposites attract" and "like marries like" are not at all at odds with each other. "Opposites attract" helps light the initial spark, while "like marries like" helps classify relationships into short—or long-term.

Handicaps

My friend and mentor Mr. Jay Conrad Levinson (author of the wildly successful *Guerrilla Marketing* series) is one of the people behind the Marlboro Man, a rugged macho cowboy who smokes cigarettes like they're going out of style. We've known for a long time that cigarettes cause all kinds of devastating health problems. So why was that ad campaign one of the most memorable and successful of all time? Why do so many ads feature people performing dangerous stunts as a way to sell everything from sports drinks to automobiles?

Peacocks and male bowerbirds provide the clues to answering these questions. In general, the peacock with the largest set of tail feathers gets the girl. These feathers can drag up to four feet (and possibly more) behind the bird, creating a display that can reach almost eight feet in width. Feathers like that are both inconvenient (imagine lugging a tail like that with you everywhere you went) and, when extended, are an open advertisement to any nearby predators. They present a serious hazard and impediment to survival. Male bowerbirds beg, borrow, and steal colored items to build elaborate nests. He with the largest, most elaborate nest gets the girl. Thievery is rampant—and why not? Stealing items from a competitor's nest enhances your chances of reproductive success at the other's expense. Here again, the time and energy invested in building and defending the bower makes it harder to find food and avoid predators.

And yet, despite these serious handicaps, the successful peacocks and bowerbirds survived well enough to find and secure mates. They triumphed over hazards greater than those of their unsuccessful peers. Overcoming that level of adversity requires a high relative degree of cunning, intellect, and physical fitness. In other words, the males who have triumphed despite facing the gravest risks probably have the best genes available, and are thus the ones most highly sought after by

Not being beautiful was the true blessing. Not being beautiful forced me to develop my inner resources. The pretty girl has a handicap to overcome.
Golda Meir

females looking for the best possible match for their own genetic material.

From peacocks and bowerbirds to stotting gazelles and chain-smoking cowboys, the pattern is clear: Those who overcome the greatest challenges get the girl. The caveat is that this "showing off" must be more than empty braggadocio; the signal must be honest, and must also come at some risk to the signaler in order to keep cheating to a minimum, lest the signal lose its value. The difference between animals and humans is that animal signals almost always tend to benefit the signaler. The wide availability of often addictive drugs and hazardous sports can make these signals extremely destructive in humans. Here is where moderation becomes important. Few people would be attracted to someone addicted to crack cocaine. On the other hand, someone who takes moderate risks might enjoy better odds of finding partners. I ride a motorcycle (with the best full-body protection I can afford), fly small planes, and take long hikes in the wilderness. I take on more risk than a good percentage of the male population, and know from personal experience that at least some women are at least somewhat attracted by that.

Economics

This is the reason why mothers are more devoted to their children than fathers: it is that they suffer more in giving them birth and are more certain that they are their own.
Aristotle

Human children are an odd bunch. Most young mammals find their own food after weaning, but humans remain dependent on their parents for several years to come. My son Logan can get food out of the kitchen and knows how to use the microwave, but he's in no position to shop for, hunt, forage for, or grow his own eats. He, like all young humans, needs his parents to feed him and show him how to start fending for himself. Women are better adapted for home life that includes nurturing children and foraging for food, while men are better adapted to providing protection and the spoils of hunts. Thanks to the uneven reproductive investment incurred by men and women, this means that a woman is best off being with a provider, but that males aren't necessarily better at being providers. Welcome back to the battle of the sexes we discussed in Chapter 14.

The kicker is that male hunters are less efficient at providing food than female foragers. Prey animals don't want to be killed and eaten, and most of them are both faster and stronger than humans. Plants and seeds are both stationary, and have no say in the matter; if anything, most of them "want" to be eaten. What hunting does provide is very concentrated calories that are very high in protein, making it anything but a wasted effort. Men also provide security against both predators and marauders. Still, they have more than a little incentive to engage in at least a little hanky panky. What's a woman to do?

Women are anything but powerless in this equation. Reproduction is the whole point of being alive. Sex is the means by which people reproduce, and both form and enhance pair bonds. Not coincidentally, women are sexually receptive throughout their entire monthly cycles, and will have even have sex during pregnancy. The book *What to Expect When You're Expecting* by Arlene Eisenberg et. al includes detailed sections describing how many months a woman can continue being sexually active during pregnancy, and how soon she can resume after giving birth.

Society provides more incentive for men to stay put and keep their end of the child-rearing bargain. Most couples live in groups that depend on each other for security and economic viability. This interdependence helps foster the type of strong pair bonds that reduce the incidence of extramarital sex, while helping keep the man around to provide for the woman and children.

The divergent interests and inherent conflicts couples (and especially couples with children) face make marriage a delicate balancing act that requires active participation by both parents to maintain. I'll discuss this in a lot more detail in *The Romantic Savage*.

There is one glaring exception to all of this pair-bonding business: Women in poverty and/or high-risk situations may choose to have as many children as possible, sometimes with as many different fathers as possible. I distinctly recall being awakened many years ago to drive a friend's neighbor to the hospital to deliver her fifth child by as many different fathers. I thought ill of that woman for a long time, only to discover later that her actions were both understandable and, by evolutionary standards, correct. Situations that carry a high risk of mortality before reproductive age (such as impoverished and/or violent areas) make having as many children a way to play

Understand that sexuality is as wide as the sea. Understand that your morality is not law. Understand that we are you. Understand that if we decide to have sex whether safe, safer, or unsafe, it is our decision and you have no rights in our lovemaking.
Derek Jarman

I've always felt that sexuality is a really slippery thing. In this day and age, it tends to get categorized and labeled, and I think labels are for food. Canned food.
Michael Stipe

the odds that one or more of them will survive. In areas where most or all men are poor providers, spreading the genetic mix may increase the odds of success. Desperate times do indeed call for desperate measures. The "shotgun" approach to reproduction scatters as many copies of one's genes as possible into the hostile environment, in the hopes that a few will hit the target of creating a new generation. Sadly, well-meaning charities (often religious) that provide food without birth control and reproductive counseling only exacerbate the situation by allowing more children to survive to reproduce, thus worsening the very problem they are trying to solve.

Lust & Sex

If men were equally at risk from this condition—if they knew their bellies might swell as if they were suffering from end-stage cirrhosis, that they would have to go nearly a year without a stiff drink, a cigarette, or even an aspirin, that they would be subject to fainting spells and unable to fight their way onto commuter trains—then I am sure that pregnancy would be classified as a sexually transmitted disease and abortions would be no more controversial than emergency appendectomies.
Barbara Ehrenreich

Most social animals have sex right out in the open, as all too many embarrassed dog owners will attest. From insects to bonobos (and with rare exceptions such as chimpanzee consortships), the animal kingdom sees public sex as no more taboo than dining at a sidewalk cafe. Let's take a closer look at what people do behind closed doors.

Concealed Ovulation

We've already discussed some of the many attraction/sexual signals humans produce, and noted our ability to have sex at any time of the month, during pregnancy, and even after menopause (which marks the end of a woman's fertility). Most of this, with the obvious exception of sex during pregnancy, is made possible thanks to concealed ovulation. Unlike other female mammals that signal ovulation (and fertility) with colors, pheromones, vocalizations, and postures (such as the peacock's fanned tail feathers). Concealing ovulation keeps everyone guessing, including women. This helps keep them receptive to sexual advances, which helps maintain pair bonds over time.

One offshoot of concealed ovulation occurs when women who work and/or live together synchronize their monthly cycles. A lone woman menstruating among a group of non-menstruating women is at a decided reproductive disadvantage. Synchronizing menstruation eliminates this problem.

One theory postulates that concealed ovulation forces the male to stay close to home, which boosts *parental confidence* (where the male knows that the children are his, and that he is not a cuckold). Another holds that concealed ovulation keeps men guessing, allowing women more choice of sexual partners. A third speculates that concealed ovulation forces the male to guard the female in order to ensure paternal confidence. Whichever theory is correct, it seems clear that obvious ovulation would cause potentially insurmountable social challenges. Imagine a bonobo-like society with all of today's technology, but without marriage or pair bonding. That may or not be a bad thing; it would certainly be different, though.

Male Equipment

Human testicles are large by gorilla standards, but only about 10% the size of chimpanzee testicles. This difference in size relates to how often the different species mate. Gorillas mate less often than humans and live in harems, meaning they have little to no competition. Humans are nominally monogamous and mate more often (thanks to concealed ovulation and pair-bonding). Chimpanzees and bonobos mate very often, and have extensive competition to be the one who actually fertilizes a female. This requires a lot of sperm, which in turn require large testicles to produce.

> *There are very few jobs that actually require a penis or vagina. All other jobs should be open to everybody.*
> Florynce R. Kennedy

Female Equipment

The female clitoris seems to have sexual pleasure as its only function. And why not? Making sex fun for females makes them more receptive to having sex, which increases the odds of having children. Human females have smaller clitorises than bonobos, which makes sense because humans don't have nearly as much sex as bonobos.

The G-spot is an area of highly sensitive tissue that lies approximately 1.5 inches inside a woman's vagina on the *anterior* (front) wall. Stimulating this area can trigger orgasms that are stronger than those available through clitoral stimulation. We'll look at orgasm in a moment. Meanwhile, it may interest you to know that human genetics are predisposed to produce female children. One man can impregnate numerous women; population growth is limited by the number of women in the population. The predisposition towards female children thus makes perfect evolutionary sense.

Orgasm

Orgasm is unusual among both apes and animals in general. Aside from the intense pleasure it brings (which induces people to keep having sex), orgasm serves several very useful purposes. The male orgasm causes sperm ejaculation. The female orgasm causes the vagina to spasm and contract in such a way that tends to draw sperm deeper inside, which also facilitates reproduction. Further, since most human sex is face-to-face with the man on top (missionary position), the woman's vagina is angled so that sperm will flow toward the uterus, thanks to gravity. The most counterproductive thing would be for the woman to get up just after sex and cause the sperm to flow out of her. Orgasms take care of this by promoting sleep and helping keep the woman relatively motionless—just what the sperm need to maximize their odds of finding and fertilizing her egg.

Love

Perfect love is rare indeed, for to be a lover will require that you continually have the subtlety of the very wise, the flexibility of the child, the sensitivity of the artist, the understanding of the philosopher, the acceptance of the saint, the tolerance of the scholar and the fortitude of the certain.
Leo Buscaglia

No talk of sex and reproduction is complete without love. I'll cover love in far more detail in *The Romantic Savage*. Meanwhile, suffice it to say that many mammals share many (if not all) of the hormones thought to be involved with love. Further, I have seen far too many examples of animal behavior in nature shows, at zoos, and in books, to where I can only conclude that animals do indeed experience love. Love keeps pair bonds alive when one parent leaves to hunt, forage, fight, etc. It helps make the inherently unequal investments in child rearing bearable, and even desirable. I can only wonder, as *When Elephants Weep* suggests, whether animals find love as joyful, complicated, frustrating, and fulfilling as humans do.

Never Enough

Ponder this: Technological advancements have elevated a prey animal to dominance over all other forms of life (assuming the human is adequately equipped). From plants to insects, fish, birds, reptiles, and mammals, nothing is immune from human activity—a fact that is becoming all too evident, as growing numbers of scientists concede that our activity is fundamentally altering the planet's climate. We have escaped predation and conquered many of the diseases that normally act to limit population numbers. Fertility clinics specialize in getting otherwise infertile people to breed, despite the seemingly obvious fact that an infertile person is that way for a reason. After all, if a mother can spontaneously abort (miscarry) an embryo or fetus, then the human body may be able to recognize potential weaknesses in itself, and thus limit or disable fertility. The "culture of life" that permeates many religions (their tolerance and/or instigation of countless wars and state-sanctioned murder notwithstanding) throws food at hungry people, while doing nothing to curb the resulting reproductive booms. Pope Benedict XVI (Joseph Ratzinger, a former precept of the Congregation for the Doctrine of the Faith—the modern name for the Inquisition) blasted Europeans for being selfish and not having enough babies. Meanwhile, the world's population mushroomed from six billion (6,000,000,000) to about six and a half billion (6,500,000,000) in only seven years. The human population grows by 3,500 people per hour, while the planet loses a plant or animal species every 20 minutes (according to the Animal Welfare Institute).

Humans have surmounted almost every natural obstacle to population growth. Our most pressing challenge is to invent new forms of controls before our population experiences the same boom and crash cycle that any artificially overpopulated species experiences. Why? Because humans are the only species capable of waging war to fight over dwindling resources, and our weapons are more than capable of destroying all life on Earth several times over.

Let me be perfectly clear: Abstinence programs don't work. Harsh penalties for premarital sex (Muslim women are jailed and executed for having sex outside of marriage, even when raped) don't work. Throwing food at the problem works temporarily, but only worsens the long-term problem. Comprehensive birth control, voluntary sterilization, and empowering women to make informed decisions about reproduction are our only hope.

The true beloveds of this world are in their lover's eyes lilacs opening, ship lights, school bells, a landscape, remembered conversations, friends, a child's Sunday, lost voices, one's favorite suit, autumn and all seasons, memory, yes, it being the earth and water of existence, memory.
Truman Capote

You can give without loving, but you cannot love without giving.
Amy Carmichael

Chapter 21

Courtship 101

> *This is courtship all the world over: the man all tongue; the woman all ears.*
> Emily Murphy

I'm willing to bet that the odds of you having just flipped this book open to this chapter are far greater than 1 in 26. (There are 26 chapters in this book.) And that's OK. Just make sure to read the other 25 chapters, to put this chapter and the chapters that follow into proper context. That said, where was I? Ah, yes...

Have you ever felt an instant "click" or connection the moment you met someone? It doesn't matter if you pass them on a sidewalk or spot them across a room. Something about that person triggered as response in you. Try walking down a crowded street sometime. As you walk, pay attention to your surroundings, but don't stare or actively look for anything or anyone in particular. There is a decent chance that you'll feel that "click" with at least one other person. What you do with that feeling is up to you; of course, you shouldn't need any reminder to remain on your best behavior at all times, but I'm reminding you anyway.

Did you feel that connection with (or attraction to) anyone you passed during your stroll? Have you ever felt something like this? If so, then you know exactly what I'm talking about. If not, well, I can all but guarantee that you'll find out exactly what I'm talking about, and I hope it happens sooner rather than later for you. I'm guessing that particular hope is mutual!

Let's talk about what triggers these moments, then move on to examine some of the ways in which people signal their attraction to each other. First, a disclaimer: I am a man writing from a male point of view, and am devoting more space to describing the woman's part of the sexual equation, because that is by far the more complex side of the story. I also freely acknowledge that individual circumstances vary widely, and that I am casting a very wide, very generic net. I cannot stress strongly enough that what follows represents absolutely no judgment on my part as to the relative worth of men, women, and/or their sexuality. As far as this author is concerned, the human species would die out without both, making both equally valid and worthy of respect and admiration. Are we clear on this? OK, then, on to the good stuff...

Courtship is to marriage, as a very witty prologue to a very dull play.
William Congreve

Initial Attraction

Peruse the personals long enough, and you'll eventually come across ads that include some variation on, "Looks not important as long as there's chemistry." I hope you remember the last two chapters, because here's a pop quiz: Was that line probably written by a man or by a woman? Why would that gender be more likely than the other to write such a line?

Attraction is beyond our will or ideas sometimes.
Juliette Binoche

If you guessed a woman, you're absolutely right. Women place far less emphasis on physical attractiveness than men. The "chemistry" women refer to is the scent given off by the man's genotype. She will probably never consciously smell the actual genotype; she will, however, pick up and imprint on the man's overall scent. Guys, here's a hint: ditch the cologne. If you're looking for a woman to fall head over heels for you, then don't mask your natural scent. Going away for a few days? Leave her one of your well-worn (not dirty or stinky, just well-worn) shirts to remember you by. You'll both be glad you did. But I digress. Women who place a lot of emphasis on physical looks are probably looking more for a fling or affair, than for a serious relationship. If you've ever seen a couple and wondered what the heck she sees in him, rest assured that she may not need to see anything, because her nose knows.

Men, on the other hand, place enormous emphasis on physical attraction. Chapter 20 lists some of the female sexual signals. Chief among these are youth, smooth skin, and facial

> *In the factory we make cosmetics; in the drugstore we sell hope.*
> Charles Revson

symmetry. Small wonder that entire industries exist to make women look more appealing. Some examples include clothing that flatters curves and/or emphasizes or de-emphasizes certain areas of the body, myriad cosmetics that mimic everything from youth to symmetry to sexual arousal, and even medical doctors, who will add/remove/modify body parts to make their patients more attractive. This has several interesting effects, including:

- Getting a makeover can render a woman almost unrecognizable to herself and others.

- Women not naturally blessed with whatever body type is currently a la mode may feel inferior and pressured to enhance their attractiveness by artificial means.

- Some women feel they need to "doll themselves up" to look good enough for a man to notice their "inner beauty."

- Women who wear too much makeup/perfume/racy clothing are often frowned upon for taking things a little too far.

- The pervasive availability and use of cosmetics and other beauty-enhancing products can place women in a self-imposed "damned if I do, damned if I don't" place. On one hand, looking good doesn't come cheap. After all, cosmetic companies and fashion designers need profits too. On the other hand, a woman who opts out of this system may find herself at a disadvantage.

Remember that women are looking for someone who can help create a child with a strong immune system. A woman can only rear a few children during her entire lifetime; she therefore has every incentive to be choosy about the man she selects. Guys, if your genotype isn't radically different than hers, cut your losses and move on; she is not, and will not ever be, that into you.

> *It isn't tying himself to one woman that a man dreads when he thinks of marrying; it's separating himself from all the other.*
> Helen Rowland

Women are also looking for a provider, someone who can and will do all he can to level out the uneven childbearing and rearing investment. Back to the personal ads: Why do you think so many of them take care to mention that the man must be employed? Why does a significant percentage of those ads specifically invite men with any "baggage" or "drama" not to apply? The former is easy. As for the latter, a man with an existing obligation to care for another woman and/or child has that much less to lavish on the new situation. I met a wonderful woman during my divorce. We clicked on every level, including the economics (see previous chapter), but it wasn't meant to be, because she was uncomfortable with the amount of support I am providing for Logan and my former wife. For the record, I absolutely refuse to be a part-time dad, and making sure my son has all he needs is my top priority. Jennifer (my current and hopefully last partner) is an MD and does not want children of her own, despite being absolutely wonderful with Logan. Her economic situation is therefore far more amenable to a relationship with someone in my position.

Many a man in love with a dimple makes the mistake of marrying the whole girl.
Stephen Leacock

Cosmetics and fashions may allow women to look younger and healthier than they really are, but don't feel too bad for the men, because just about any man can don the trappings of an excellent provider. He can scrimp and save for a power suit that will make him the spitting image of a wealthy magnate. Too much work? No problem. A vast assortment of shortcuts can transform all but the most indigent among us into millionaire look-alikes. Thanks to creative financing arrangements like charge cards, personal loans, auto leases, luxury condo rentals, layaway plans, same-as-cash deals, and more, the enterprising male can look positively well off... at least until the bills start arriving. There is a very fine line between bolstering one's image and lying about it. Putting a fresh coat of paint on a solid house before selling it is a nice touch. Painting over dry rot, termite damage, cracks, etc. is both not nice and may get you in very serious trouble. I trust the analogy is clear.

To sum up, a woman's initial attraction to a man will probably be biochemical (genotype) and physical (to the extent that the man looks like he could be a decent provider and that he's both clean and clean cut).. Gentlemen, be clean and well groomed at all times, and Nature will take its course. A man's initial attraction to a woman will probably be physical. Ladies, it's OK to enhance yourself. Just don't go overboard. The right man will think you're hot stuff without all the fancy (and hopefully not animal tested) cosmetics.

I want to keep my attractiveness as long as I can. It has to do with vitality and energy and interest.
Jacqueline Bisset

Attraction Signals

> *Sex is a part of nature. I go along with nature.*
> Marilyn Monroe

My friend Shannon once remarked that it is impossible for people to be in mixed company without everyone thinking about having sex with everyone of the opposite (and possibly even the same) gender. If you're a man and you've ever shared a space with one or more women (even for a few seconds), they have all thought about having sex with you, and you have thought about having sex with all of them. It does not matter if the others in question are total strangers or people you've known your entire life. These thoughts go through everyone's head in mere fractions of a second, and are almost always completely subconscious. The mere presence of these thoughts does not imply attraction; a close relative will trigger the thought, followed by an immediate "no, thanks." These thoughts can become conscious if, for example, your nose detects a particularly appealing genotype, or your eye catches a particularly comely individual. They can also become conscious if you know to expect them, as you now do after reading this paragraph. Observe others around you closely enough, and you may just be able to tell that the same thoughts are running through their minds. Why does this happen? Because each of us is alive to reproduce, and we are therefore always on the make, whether we know it or not, and whether or not we want to admit it.

> *Confidence is at the root of so many attractive qualities, a sense of humor, a sense of style, a willingness to be who you are no matter what anyone else might think or say and it's true, I do have a certain fondness for women that have dark hair.*
> Wentworth Miller

Fleeting subconscious thoughts are one thing; attraction is quite another. Here again, it is all but impossible for someone to contain their attraction to another person, because doing so requires a level of concentration and acting ability that only a few people possess. In fact, the more someone tries to hide their attraction, the more obvious the remaining clues will probably become to the astute observer. Someone trying to hide her or his attraction to someone else will radiate vibes analogous to "don't think of the color red," which is precisely what you just thought of. In this case, though, the intended signal (on a conscious level) is "Don't for a moment believe that I am attracted to <insert name here>."

Here are a handful of the great many ways in which people let their attraction to others slip:

- *Body posture.* The person may pose so as to prominently display a part of her or his body, such as sitting in such a way as to let a skirt show more leg, flexing muscles, looking directly at the object of their attraction, stiffening or freezing (think deer in headlights), pouting, etc.

- *Behavior.* The person may smile or otherwise perk up as their heartthrob approaches. At the other end of the spectrum, challenging or otherwise appearing to stand up to or brush off the same heartthrob may indicate attraction. Laughter or other emotions may seem forced or excessive. The person may appear flustered, distracted, uncoordinated, helpless, etc. especially when near the object of their affection.

- *Movement.* An exaggerated walk (such as strutting or swinging the hips) is a good indicator of attraction. Subtle touches (a hand on the arm while laughing at a joke, for example), actively trying to remain close by, compulsive grooming (flicking hair, applying lipstick, checking and re-checking one's tie), and more are also possible signs of attraction.

- *Specific attention-getting behavior.* The person may actively signal their attraction in many ways, such as acting helpless (a dead giveaway when the actor or actress is normally a model of self reliance), making some kind of loud noise (verbal or otherwise), making a comment that begs a reply, acting giddy or playful, or avoiding eye contact with the object of attraction.

- *Direct invitation.* The movie *Hitch* was absolutely correct: If a woman plays with her keys when getting dropped off, she's not ready for the evening to end. Did she wave an empty glass (probably as part of a friendly "hello")? Did he (or she) flat out state her or his feelings and/or intentions?

This list is neither all-inclusive, gender specific (substitute any gender-specific pronoun you like in the above list), nor is it absolute by any means. It should, however, provide enough clues to help you figure out when someone is attracted to you, and to break down any self-denial you may attempt should you ever find yourself drawn to someone. I think it all comes down to something another friend told me a long time ago: If someone starts acting just plain goofy around you, they probably have at least a small crush on you. And if you're attracted

---

Seduction is always more singular and sublime than sex and it commands the higher price.
Jean Baudrillard

You can seduce a man without taking anything off, without even touching him.
Rae Dawn Chong

to someone else? Don't worry; they probably figured it out long before you did.

Consummation

Chastity—the most unnatural of all the sexual perversions.
Aldous Huxley

I hate to be the bearer of bad news, but picking someone up at a bar is generally a poor strategy. I am the first to point out that there are enough exceptions to this rule to keep plenty of watering holes open throughout the world, but the fact remains that the overall odds are fairly low. In fact, it is the rare man who will find a woman who is willing to have sex with him immediately or almost immediately. A prospective male must usually wait at least a week to have any real chances of success. His odds increase dramatically between one week and one month of meeting the woman and keep increasing from there at a much slower rate up to the one year mark, after which they decline. Bottom line: A man who meets a woman and waits a month will have the best chance of success with a given woman that he'll probably ever have. Wait more than a year, and the chances will go right back down.

Why? Remember that male attraction tends to be physical. It only takes a second to spot youth, facial symmetry, middle finger length, and other physical traits to arrive at a yes or no answer to the "do I like her?" question. This decision needs only lust; the male has every incentive to impregnate the woman, and then either stay to help raise the child or go off in search of new conquests. This is why 75% of men would agree to have sex with a woman he just met.

I have always felt the basis of everything in life is sexual, and I will maintain that to my dying day.
Frank Langella

Almost no woman would have sex with a man she just met, no matter how different his genotype is or how good a provider he seems to be. Women are fully aware of men's tendencies towards promiscuity, even if they aren't aware of the evolutionary reasons behind it. They are also fully aware of the uneven reproductive investments between the species and their own very limited lifetime reproductive potential. Looking out for their own interests is of prime importance to them, and they need time to get to know the man somewhat to see if the reality lives up to the hype. Our primordial past with its distinct lack of credit and other means to fake status and ability made one to four weeks plenty of time to observe and make decisions. Again, civilization has evolved too rapidly for our primordially-minded brains to keep up. This may explain why the chances of a woman accepting a man for sex keep increasing for the first year. At a certain point, however, she or the man will need to proverbially either fish, cut bait, or get out of the boat—hence the declining odds after the first year. Guys, any woman you meet needs time to size you up, to see if pyschographic/demographic, economic, and perhaps a little hazard can match up to the initial biochemical spark. You have everything to gain—and nothing to lose—by waiting and appreciating the woman for who she is, without ever once trying to press the issue.

I think for most of us, our biggest frailties are sexual.
George Michael

I am not trying to paint either gender as more into dating, relationships, or sex than the other. 50% of men and women would agree to date a stranger. There are plenty of single men and women out there actively looking for partners. All I am saying is that each gender is genetically programmed to watch out for its own interests, because no person is here to pass along someone else's genes; we are all here to pass along our own genes, and must employ our own criteria for determining who to enlist to help us in that endeavor. This is why less than 10% of women will visit a man's home on a first date, while about 70% of men would visit a woman's home on a first date.

Pre-Copulation

Much of the initial pair bonding occurs in public, both because most budding couples want to go do fun things together, and because of the woman's (usually) natural reluctance to go anyplace private before having a chance to appraise her new potential mate. The worst thing a man can do at this point is push; in fact, showing some disinterest or indifference may score him some serious points. Women are choosy, and rightly so, for all of the reasons presented in this book, and then some. A woman's very anatomy forces her to stop and consider before having sex, because the pain of breaking the hymen is not something to take too lightly. The hymen is evolution's insurance policy.

It is love rather than sexual lust or unbridled sexuality if, in addition to the need or want involved, there is also some impulse to give pleasure to the persons thus loved and not merely to use them for our own selfish pleasure.
Mortimer Adler

During this time, the man is probably focusing on physical allure, while the woman is concentrating on the emotional side of the equation because, again, she is searching for a decent provider. It doesn't matter whether she has any conscious intention to have children or not; the same evolutionary instincts are at play and are guiding her every move, whether or not she's even aware of them, and whether she likes it or not. The exceptions are where the woman is clearly looking for a completely casual arrangement, and/or if she's looking to get pregnant by any means and with anyone possible (because some chance at passing on genes beats no chance). Men are no less shaves to their evolved instincts. I am focusing on the woman, because it is ultimately her choice whether or not to have sex. Books like *The Game* by Neil Strauss offer methods to short-circuit a woman's natural protective instincts—good for a one-night stand, perhaps, but don't try to build a lasting relationship using false pretenses.

As the relationship progresses, the pair-bonding will become increasingly intimate and increasingly private.

Copulation

My classmates would copulate with anything that moved, but I never saw any reason to limit myself.
Emo Phillips

Some of the many unique aspects of human sexuality involve our vertical posture (hard to see genital swellings or coloration that way) that led to vertically-oriented sexual signals. The most prominent of these signals, the woman's breasts, are analogous to her buttocks. So why do they face forward when a woman's buttocks are behind her? There are several reasons. First of all, having both breasts and buttocks makes a woman stand out whether she is coming or going. The combination of breasts and buttocks ensures that a women is transmitting sexual signals across the full 360-degree circle surrounding her, and across at least 180 degrees when she is sitting or lying down. It's simple: The woman attracts interested men and then makes her selection, just like so many animal species from apes to birds, reptiles, and amphibians.

The strong pair bond required to grow and maintain a long-term pair bond suitable for raising children makes face-to-face copulation ideal (as opposed to other positions, such as the male kneeling behind a woman on all fours, which is the standard mating position of a great number of animals). Rear entry certainly gets the job done; it also minimizes the contact between the partners to the minimum needed to do the deed. Face to face copulation is an intensely personalized experience that strengthens the pair bond. In this situation, a woman's breasts are perfectly positioned to be in full view of the male throughout the entire process. About 70% of human copulation occurs face to face, most often in the so-called "missionary" position with the man atop the woman (leaving the protector more able to respond to unforeseen threats while protecting the woman). Our physical anatomy is designed to facilitate this position, as I described in Chapter 20.

Post-Copulation

Chapter 20 also explained the benefits of female orgasm from a fertilization point of view. An important additional benefit of female orgasm is that it feels good, which rewards the woman for having sex and gives her incentive to do it again sometime. There is another possible benefit as well: The female orgasm is an elusive beast. Comedian Robin Williams joked that getting a woman to climax is like assembling a jigsaw puzzle whose pieces keep changing. There's a very good reason for this that goes right back to the reality of women looking for a provider. Providing for someone on an ongoing basis requires paying a lot of attention and responding to ever-changing circumstances. Men who are better able to do this have better odds of keeping the woman interested in him. Make no mistake: The appraisal process I've been describing is ongoing. It never ends, especially the first few times the couple has sex.

Orgasm: the genitals sneezing.
Mason Cooley

Male orgasm serves several purposes as well. It inseminates the woman (the whole point of the exercise) and also feels good (as if the man needed more incentive, right?) It also causes him to feel drowsy, a fact that also forms a stock in trade for any comedian or comedienne worth the title. Guess what: This keeps the man near the woman and the refractory period (the recovery period after an orgasm when it's difficult or impossible for a man to have sex again) ensures that he won't go off philandering anytime in the immediate future. The woman, who is obeying her evolutionarily programmed imperative to lie still, remains protected long enough for fertilization to take place.

Keeping the Faith

Mama's baby, Papa's maybe. This old saying goes to the core of male/female relationships and the issue of fidelity. In gen-

People forget we come from an embryo and we're part sperm and part ovary. We have both sides in us.
Michelle Rodriguez

> *It is a wise father that knows his own child.*
> William Shakespeare

eral, a woman can be 100% certain that her child is indeed hers, because it's impossible to give birth to someone else's baby (with the notable exception of surrogate mothers). The father has no such assurances. Witnessing the baby being born and knowing that he had sex with the mother nine months prior is no proof that he is the actual father. This concern is anything but trivial; a cuckold is evolution's biggest loser because he is investing his precious resources to help the real (in the biological sense) father pass along his genes. Adoptive parents like me also fall into this category; however, unlike cuckolds, we enter this to arrangement deliberately, and at least get to pass along the nurture side of the nature/nurture equation. I don't feel like any less of a father because Logan doesn't carry my genes.

Fathers have good reason to worry: A study conducted in England during the 1940s revealed that nearly 10% of babies are the result of adulterous encounters, meaning that 1 in 10 British men are cuckolds. The actual percentage of women committing adultery must be much higher, because most liaisons don't result in pregnancy thanks to contraceptives and the fact that a woman's peak fertility only spans about three days every month. Lest any man reading this feel morally superior, remember that each of these women have male partners. Subsequent studies of British and American babies using ever more sophisticated techniques indicate that between 5% and 30% of babies born in these nations were adulterously conceived. Think about 20 fathers you know. At least one and as many as six of them are cuckolds. Do the real fathers know this? Probably not. Do the women? That depends on individual circumstances. If the woman had sex with her husband around the same time she had the affair, she may believe that the child is her husband's. On the other hand, a woman giving birth to a baby of a different ethnicity is pretty hard to hide. (This happened to a friend of a friend of mine some years ago.)

> *Adultery is the application of democracy to love.*
> H. L. Mencken

This explains why laws and mores dealing with adultery tend to regulate women more than men. Male fears about paternity prompts them to restrict their wives opportunities to have sex with other men. The more extreme examples of this kind of regulation include:

- *Virgin premiums.* Plenty of men in plenty of cultures value virginal women, and even go to such lengths as hanging the bloody sheets for all to see the morning after their wedding night and consummating the relationship.

- *Clitoral castration.* Removing the woman's clitoris (sometimes under less-than-sterile conditions) is thought to reduce the woman's sex drive and the enjoyment she receives from sex.

- *Infibulation.* Some cultures sew girls' vaginas shut, leaving a small opening to allow menstrual blood to pass. Some husbands consummate their marriage and remove the stitches in one motion by simply penetrating the woman as if there was nothing in the way. I can only imagine the pain that causes for the woman!

- *Chastity belts.* Metal devices with small (usually barbed) openings to allow excretion and menstruation. These are locked in place and the husband keeps the key.

- *Extreme punishments.* Recent news reports have told of women jailed for being raped (presumably the woman brought it on herself by tempting the rapist), and of other women being executed by stoning, or by being doused with gasoline and set ablaze, for committing adultery or having premarital sex. Muslim men who commit adultery with non-Muslim women incur no penalty, as far as I can tell. Muslims have no monopoly on such punishments. The book of Leviticus specifies that adulterers of both sexes must be put to death. And lest any Christians think Jesus was soft on adulterers, the book of Matthew specifies that someone who has lusted after another has committed adultery, as has someone who has divorced and then remarried. Meanwhile, men were free to have multiple wives, but women could have only one husband.

- *Restrictive clothing.* Burqas and other clothes designed to conceal a woman's body and curves are required wearing in certain parts of the world. Some catalogs feature swimwear that looks like Victorian-era garb, except that modern synthetic fabrics make it possible to swim with some freedom of mobility (and without being pulled under by the outfit's sheer weight). Again, lest anyone

---

It is an infantile superstition of the human spirit that virginity would be thought a virtue and not the barrier that separates ignorance from knowledge.
Voltaire

We may eventually come to realize that chastity is no more a virtue than malnutrition.
Alex Comfort

think I'm picking on Muslims, the Victorians practically invented cumbersome, all-over "swimsuits" that no modern woman would wear in the rain, much less while trying to actually swim.

> *I lost my virginity when I was 14. And I haven't been able to find it.*
> David Duchovny

The list could go on and on, but I think the point is clear: Men are obsessed with virginity and chastity, which is an unavoidable consequence of the way human sexuality evolved. Our laws and customs evolved to ease these fears, which had the effect of removing sex from public purview and making relationships exclusive.

If you don't believe me, fill in the blanks to complete some of the following popular expressions made about infants:

- "He looks just like his _____."
- "She has her _____'s eyes."

If you guessed "father" for both examples, you're absolutely correct. Remember that staying around to raise children is only in the man's best interests to a certain extent. Taking pins to ensure paternal confidence provides powerful added incentive for the man to go the extra mile for his wife and children, which in turn benefits the wife. The upshot of taking measures to ensure paternal confidence is that more men invested more in raising their children. The human population continues to boom today and grew by about 20,000 people in the time it took me to draft this chapter—the equivalent of adding another city of Ashland, Oregon to the planet.

Games People Play

> *Most games are lost, not won.*
> Casey Stengel

It should be blatantly obvious by now that human sexuality and courtship is rife with game playing on all sides. I think it no exaggeration to state that all relationships include at least some game playing by and among both partners, an observation based on personal experience and observation. I will even go so far as to say that a relationship without any obvious games has said lack of obvious games as its primary game. If you find yourself more than a little confused and baffled at this point, well, welcome to the human condition. Thousands of years of philosophy, religion, law, psychology, psychiatry, sociology, and more have come no closer to resolving the battle of the sexes or to eliminating the games men and women play in their relationships. Many of these games are harmless; flirting falls into this category. Other games are extremely serious, such as a woman who falsely accuses a man of raping her because he's not interested in her or her advances. Men are equally guilty of playing their own games. Let's take a look at some of the most common games people play and why they play them.

- *Keeping others waiting* (the VIP). This often happens because the person playing the game doesn't want to seem too interested and push the other way by seeming too eager.

- *Entrapment* (the bad cop). The player asks a question that has no correct answer. One answer leads to a confrontation ("You think I'm fat!"), while the other leads to accusations of dishonesty ("You're just saying that.").

- *Playing ignorant* (the stereotypical—and gender neutral—dumb blonde impression). It's amazing how dumb/amnesic someone can become when confronted with something he or she just doesn't feel like doing.

- *Feigning disinterest* (the above-it-all celebrity). Someone giving off attraction signals (see above) while pretending not to notice, or even not to like the object of their affection, is playing a game designed to see if the other will pick up on the cues and take the bait.

- *Testing priorities* (the secret agent). Person A is deep into something, and Person B starts a conversation or catalyzes some issue to grab attention and test the other person's priorities. This is a risky one, because ignoring the attention is rude, but dropping everything to stop and deal with it is a surrender of power that won't soon be forgotten.

- *Trading services* (the commodities broker). Women know that men have sex on the brain at all times, and will use that to great advantage. Men: Think about some of the things you've done with the promise of having sex that evening. Women: Think about some of the times you've played that game. Now reverse the roles, because men will most certainly initiate this game, promising to take

I think I'm being friendly with someone and I'll sit in their lap. They think I'm flirting with them.
Kylie Minogue

Sometimes I think I might insult people by being openly flirtatious, then snatching it back.
Dominique Swain

the woman shopping (for example) if she'll have sex with him.

- *Guessing games* (the charades game). Forget coming out and saying what's on one's mind. Why not instead drop subtle cryptic hints that are virtually guaranteed to fail only to take out your anger with yourself on your unsuspecting partner? This is a great way to keep a partner on her or his toes, because s/he will parse what you say to try to determine if you really meant what you said, if you meant the opposite, something in between, or something totally unrelated. Just don't expect them to get it right.

- *Cold shoulder* (the aloof approach). This is a close corollary to the guessing game, except that the player maintains a tense silence or otherwise ignores the other's presence.

- *Ultimatums* (it's good to be the king/queen). Do this, or the relationship is over. At this point, the question becomes whether or not this game is a bluff, and whether or not you want to call that bluff.

Time for another pop quiz: How many of the above nine games have you played in your lifetime? Think carefully; some of your best games may not seem like games at first glance. Forget about justifications and don't get defensive. All you need to do is be honest and think of the correct number.

All set?

Give yourself one point for every one of the games you've played from the above list. Add them up and score your results:

- 1-8. You're either kidding yourself, or you're not human.

- 9: Congratulations! You're a living, breathing member of *homo sapiens*.

Relationship games are a fact of life brought on by the very uneven parenting investments and competing priorities between the genders. Getting around that fact of life requires nothing less than completely reprogramming human sexuality, an endeavor that would simply replace the games we play today with a new set (such as trying to put on the best-looking

---

You know, I'm not really any good at working out when people are flirting with me. And I think I'm too flirtatious with people I'm trying not to flirt with! What I am good at is making people feel uncomfortable. I don't want to but it always ends up happening!
Josh Hartnett

tail feather display, or build the biggest bower while fighting off would-be thieves). It's all part of the built-in conflict that lies at the core of what it means to be alive.

In Summary

The best evidence available seems to point to human nature that is designed to get us into relationships that last long enough to raise the next generation to biological adulthood, a sort of serial monogamy if you will. Humans have taken this evolutionary ball and run with it to say that people should marry for life and forever perish the thought of so much as fantasizing about anybody else. I believe it is physically impossible to stop sexual fantasies altogether, and am just cynical enough to believe that religions use this to their advantage to keep people fearful and subjugated. More on this in Chapter 23.

I believe it's both inevitable and only fair that any reproductive system that involved uneven parental investments will result in the gender shouldering the greatest burden to be as choosy as possible, which forces the other gender to do all it can to at least give the impression of doing and/or being whatever it takes, in order to have that precious chance at passing on the genes. The game board is set, and all that's needed are the players. The catch is that no one can afford to make their games too obvious for fear of being seen as dishonest, nor can they make any conceding moves very obvious, for fear of being seen as a pushover. And around and around we go...

The reproduction of mankind is a great marvel and mystery. Had God consulted me in the matter, I should have advised him to continue the generation of the species by fashioning them out of clay.
Martin Luther

Chapter 22

Neither Straight nor Narrow

Politics is my hobby. Smut is my vocation.
Larry Flynt

Smut, if it's really smut, there's nothing backing it up. It's the easy way out.
Sandra Bernhard

Sex is one of the most misunderstood and maligned aspects of humanity. I think it's a crying shame that nations such as the United States condone violence, while having a collective tizzy over something as trivial as a nipple briefly exposed during a halftime show. The Puritanical beliefs that founded this nation are alive and well, and are even enjoying a resurgence, thanks to the growing might of the conservative evangelical Christian movement. Other cultures and religious movements around the world follow the same pattern of believing that sex is something dirty that needs strict regulation. On the other hand, Europe and other parts of the world seem perfectly content to eschew violence and at least accept (if not downright celebrate) human sexuality. And why not? All human cultures are steeped in sex, whether they choose to admit it or not. The Europeans are just being honest.

If we assume that the average human sexual encounter lasts just one second from start to finish and that each couple has sex only once per year, then about 77 couples are having sex somewhere on this planet at any given second. Consider that the average person has dozens of sexual encounters per year that each last far longer than one second, and the number of people having sex at any given moment skyrockets.

Set aside whatever moral or other convictions you have about sex for a few moments as we look at some statistics (as reported by the Kinsey Institute from a number of studies from various sources). As you read these numbers, consider them against all we have covered so far in this book, and then realize that no amount of attempting to squash or inhibit sexuality can ever hope to succeed. When it comes right down to it, we are bound to follow our genetic programming. As Dr. Ian Malcolm from the movie *Jurassic Park* (played by actor Jeff Goldblum) explains:

> *"If there is one thing the history of evolution has taught us, it's that life will not be contained. Life breaks free, expands to new territory, and crashes through barriers, painfully, maybe even dangerously."*

I couldn't have said it better myself.

Sex on the Brain

One urban legend claims that people think about sex every seven seconds. That's not quite accurate; about 1 in 5 women think about sex at least once per day while, 2/3 think about it at least a few times per week or month. It should come as no surprise that the numbers for men are noticeably higher. I am confident that these numbers could go up or down, depending on how one defines "thinking about sex," and what specifically constitutes sex (coitus? making out? kissing?). Any way you look at it, humans think about sex a lot. This is just what one might expect from a species whose sole evolutionary purpose in life is to reproduces.

From the moment I was six I felt sexy. And let me tell you it was hell, sheer hell, waiting to do something about it.
Bette Davis

Thoughts are one thing, but action is quite another. So, when it comes to sex, do our deeds match our thoughts? Young adults under 30 have sex more than 100 times per year on average, a frequency that drops to around 69 times per year during their 40s. You be the judge. Chimpanzees may mate more often than we do, but never forget that Earth's three hominid species (*homo sapiens*, *pan paniscus*, and *pan troglodyte*) are all very much the exception among mammals and animals in general. Each of these species has a lot of sex!

Sex in Practice

> *Speaking for myself, my very integrity as a human being needs to include my freedom to explore who I am both spiritually and sexually. Not just to explore—but to practice.*
> Malcolm Boyd

The beginning of this chapter asserts that someone somewhere on Earth is having sex at any given moment of any given day. In fact, chances are excellent that many thousands of people are having sex at any given time. The Kinsey Institute reports that 90% of men and 86% of women have had sex in the past year. The US Census Bureau predicted a total world population of 6,679,493,893 by the middle of 2008, of which 4,866,736,385 were 15 years of age or older. Let's assume that this population is equally divided among males and females, and that every couple has sex just once all year. We'll also round all fractions down to the nearest whole number. This gives us 2,433,368,192 couples, of which an average of 6,666,762 will have sex on any given day, which translates to 277,781 couples per hour, 4,629 per minute, and 77 per second. Compare that to 1.8 deaths per second (US Census Bureau projections for 2008). That's a lot of sex, even if we don't factor in real numbers regarding actual frequency and duration of sexual encounters. Taking real numbers into account yields far higher numbers; the World Health Organization estimates that 100 million acts of sexual intercourse occur every single day!

My point is simple: Sex is all around us, whether we like it or not. We may as well enjoy sex because it sure beats some alternatives!

Frequency

> *Very few people can truly divorce themselves from what they feel emotionally and sexually.*
> Boy George

About three quarters of non-married people claim to have had zero to only occasional sex during the past year, while about 6% claim to have sex four or more times per week. For married people, about a third claim to have had zero to only occasional sex during the past year, while about 7% claim four or more times per week. Being married is apparently good for a person's sex life, which is absolutely predictable given what we've discussed so far about human sexuality. Married couples have every reason to have sex, both to procreate and to maintain the pair bond necessary to fully reap the fruits of those labors in the form of a viable genetic legacy.

How do these numbers correspond to the average marriage durations listed in Chapter 20, and what should one make of the persistent reports about waning sexual frequency over the course of a marriage? The best conclusion seems to be that married people have a lot of sex at the beginning, but then taper off over time, again the exact behavior one might expect to see in an animal whose chief goal in life is to reproduce, and whose marriages tend to last long enough to see the children to puberty but not longer.

Orgasm

The good news is that 75% of men always have an orgasm with their partner. The bad news is that only 29% of women enjoy this same pleasure. Still, more than 60% of women report being very happy with both the quantity and quality of their orgasms. Many women also say that their best sexual experiences are those where they feel connected to their partners, as opposed to simply having orgasms. This feeling of connection probably provides some reassurance that the man will stick around to provide for her and the children, and its importance is therefore easily understood. Four out of every 10 sexually active adults expressed extreme satisfaction with both the physical (sexual) and emotional (pair bond) aspects of their relationships.

The orgasm is simply when the body does take over.
Betty Dodson

Simultaneous orgasm is a wonderful thing, both for the participants and on an evolutionary basis, because that means that the woman's body is actively drawing sperm in at the exact moment that the man's body is actively providing that sperm. 25% of men and 14% of women consider simultaneous orgasm essential. That a higher percentage of men feel this way seems right in line with the male bias towards physical performance and status. That fewer women share this opinion seems to corroborate the high relative importance of emotional connections, as opposed to the merely physical.

Starting Young

Seems like every generation has its fun, only to admonish the next generation against doing the exact same thing. Curfews, crackdowns on teen/young adult partying, conservative morals, abstinence programs, dress codes, and other measures attempt to douse or at least subdue raging hormones. Do

Youth has no age.
Pablo Picasso

> *Passion rebuilds the world for the youth. It makes all things alive and significant.*
> Ralph Waldo Emerson

these measures work? Fully one in four males have had sex by the age of 15. By age 24, that number jumps to just about nine in ten. Those who think that boys will be boys while young girls are remaining chaste are in for a rude shock, because 26% of females have had sex by age 15, a number that climbs to 92% by age 24. Girls are more likely to have had sex at any age between 15 and 24 than men. The downside is that one in ten females becomes pregnant by age 19, which represents 19% of all sexually active women in this age group.

These numbers speak for themselves. The whole point of living is to procreate, and a teen/young adult body is as fit and prime for reproduction as it will ever be. Having sex earlier means having babies earlier, which gives the parents the best chance of having as many children as possible in their lifetimes. Keep in mind that modern civilization and the tremendous drop in both infant mortality and child mortality due to such things as predation, injury, illness, or starvation has occurred far too recently for our brains to evolve to this new reality. Our minds are still programmed for life on the African savannah, where having many children is absolutely essential if one is to have any shot at a genetic legacy. Modern developments, such as medicine, and overpopulation that make it both safe and prudent to only have 2 children (to replace the parents), simply have not made it into our genetic makeup.

> *The big mistake that men make is that when they turn thirteen or fourteen and all of a sudden they've reached puberty, they believe that they like women. Actually, you're just horny.*
> George William Curtis

From an evolutionary standpoint, hitting sexual maturity without going out and having as much sex as possible is like idling a car—a waste of precious time and resources. Gas tanks can be refilled. One's reproductive potential can never be replenished. All of the taboos, and other restrictions a society attempts to place on sexuality, will be ignored and/or circumvented. There are nations today where someone caught having sex outside of marriage risks being brutally murdered, sometimes by large mobs of people. The fact that people continue to suffer burning, stoning, imprisonment, and other extreme penalties betrays the utter ineffectiveness of these measures, because the need to reproduce both early and often is hard wired into our most fundamental building blocks. It is no coincidence that starving monkeys will choose sex over food.

The urge to procreate will overcome all obstacles. The sooner people of all religious and/or political beliefs realize this, the better off we will all be. Modern contraception offers extremely reliable protection against unwanted pregnancies, especially when multiple methods are used in concert (such as a man wearing a condom while the female uses a diaphragm and spermicide). Extremely successful public service campaigns have persuaded people to wear seat belts and take other basic precautions that have saved countless lives. Expanding "sex ed" from a half-semester course taken once during high school to an ongoing course taught throughout both middle and high school, with a strong emphasis on contraception and other safe sex practices, would go a long way toward curbing both teen pregnancy and the spread of sexually transmitted diseases. Other methods have been tried, and all have failed miserably, results that anyone familiar with basic biology should find utterly predictable.

Contraception

A 2002 study (AGI) of women aged 15-44 found that 62% of these 62 million women use contraceptives, with the Pill being the most common option used by just over 30% of them. Tubal ligation surgery ranks second, with the male condom coming in third (18%). I should note that, of these methods, only the male condom offers any protection against sexually transmitted diseases.

A study conducted by Mathmatica Policy Research in 2007 found that "abstinence only" sexual education had zero (zero, zip, nada, zilch) effect on teen sexual activity or the age at which teens began having sex. Furthermore, teens participating in abstinence programs were less likely to use condoms (to protect against disease) than teens not involved in these programs, despite being more aware of their existence than non-participating teens. Well, let's see... A teen buying condoms is announcing her or his sexual activity. Being sexually active while enrolled in a program designed to maintain celibacy may well induce guilty feelings and/or a perceived need for secrecy, thanks to the religious and/or moral themes that lie at the root of these programs. Far from eliminating or even reducing teenage sexuality, abstinence programs are simply driving it underground and making it more dangerous. Remember that the threat of violent death is not enough to stop young people from having sex, making it either wishful thinking (best case)

Despite the availability of cheap and effective contraception, it looks as if we are not as careful in our decisions about reproduction as all the talk of family planning might suggest.
Hugh Mackay

or deliberate self-delusion (worst case) to think that the far lower stakes faced by teens in such programs will have any impact whatsoever (aside from driving the activity underground, of course).

Serial Monogamy

> *I think that monogamy is artificial. I do not think it's something that comes naturally to us.*
> Tom Ford

How do the statistics we just saw about sexual activity by age correlate with the statement in Chapter 20 about the average marriage lasting between eight and ten years? This is an excellent question. Men between the ages of 30 and 44 have had an average total of 6-8 sexual partners with women in the same age range averaging four.

The following chart displays the percentage of men and women aged 30-44 who report having had differing numbers of sexual partners in their lives since age 18:

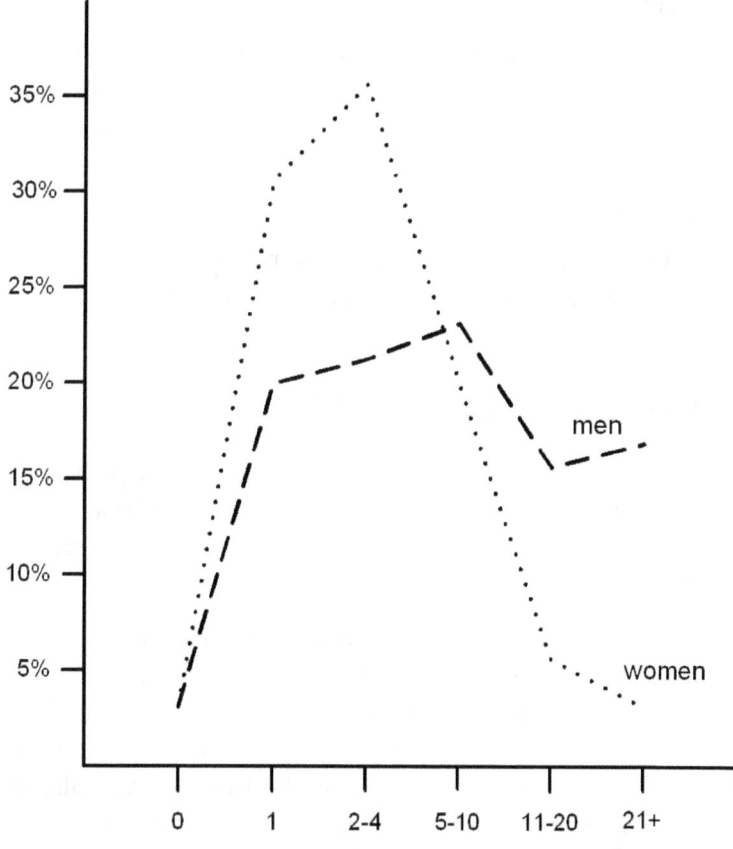

> *I think serial monogamy says it all.*
> Tracey Ullman

As you can see, these percentages are absolutely consistent with the theory that humans are serially monogamous at best while pursuing a mixed reproductive strategy, where the need to be faithful and raise the young competes with the need to simply sire as many young as possible (a dilemma faced by men far more often than women as we've discussed, and as this chart attests).

Notice the great disparity in reported numbers of sexual partners among men and women, where far more men report having higher numbers of sexual partners, while far more women report having fewer sexual partners. 56% of men report having five or more sexual partners in their lifetime, compared to only 30% of women. There are several possible reasons for this, including:

- Men may be more likely to report greater numbers of partners, since this is generally seen as a status symbol. Men who are thought to be promiscuous are admired and generally referred to as studs, players, ladies' men, etc.

- Women may be more less likely to report greater numbers of partners, since this is generally frowned upon. Women who are thought to be promiscuous are labeled whores, tramps, floozies, sluts, tarts, and a host of other, equally denigrating terms.

- The most promiscuous women may have far more sexual partners than the most promiscuous men.

Hard work is damn near as overrated as monogamy.
Huey Long

Beyond Vanilla

OK, so people have lots of sex with lots of other people. At least the vast majority of this sex consists of the basics (missionary position) with the "freaky" stuff kept to a distinct minimum, right?

Wrong. When it comes to inventiveness, variety, and just plain kink, imagination is the only limit to the sexual games people play. It is no accident that the sex industry pioneered the World Wide Web and developed countless innovative technologies for delivering adult content on the Internet. You already know that many of the appliances we use around the house (such as microwave ovens) got their start in the military. What you probably don't know is that much of the technology we

I also use women as a sex object; maybe I'm kinky. However, I like to talk to them as well.
Oliver Reed

take for granted on the Internet (streaming video, paid members-only sites, e-commerce, etc.) were either invented or heavily developed and advanced by the online sex industry.

Fantasy

Dreams are not without meaning wherever they may come from—from fantasy, from the elements, or from other inspiration.
Paracelsus

It should be no surprise to anyone that male sexual fantasies are far more explicit than female fantasies, which tend to center around emotional connections and romance. Men also fantasize about the high sex drive of their ideal partners and the sheer pleasure of the sexual acts themselves to a far greater extent than women. What else can one expect from an animal built from the ground up to spread his seed, especially when said animal has none of the investments of monthly cycles, pregnancy, nursing, etc. to contend with? Here yet again, numbers bear out the assumptions: Over half of men think about sex at least once per day while 43% of them think about sex anywhere from a few times per month to a few times per week, with less than 1 in 20 thinking about sex less than once per month.

Masturbation

I was an altar boy, a spokesperson for the Virgin Mary, I was a choir boy but then at the age of 14 I discovered masturbation and all that went out the window.
Guillermo del Toro

98% of undergraduate college men and 44% of their female counterparts admitted to ever having masturbated, with the men doing it about 12 times per month, the women 4.7 times. 60% of men and 40% of women admitted to masturbating during the prior year. One might think these numbers would drop among people living with significant others, but the opposite is in fact true: about 85% of men and 45% of women who were living with their sexual partners masturbated in the past year. Over half of males and 25% of women masturbate for the first time between the ages of 11 and 13. And finally, about 5% of men and 11% of women have never masturbated (or so they say).

Pornography

Erotica is using a feather, pornography is using the whole chicken.
Isabel Allende

Forget books, electronics, and auctions. When it comes to electronic (Internet) commerce, the sex industry accounts for over half of all online spending. By some accounts, the total combined revenue earned by all pornographic Web sites in the United States alone exceeds the combined income of broadcasting giants ABC, NBC, and CBS. One quarter of men and 4% of women have visited at least one pornographic Web site in the last 30 days. As for "offline" or "traditional" sales, 23% of men and 11% of women have purchased adult movies or videos. This does not include other sexually related material such as toys, lingerie, etc.

Oral Sex

A 1993 study conducted by Janus & Janus found that 10% of men and 18% of women prefer oral sex for achieving orgasm. 27% of men and 19% of women have had oral sex within the past year.

Anal Sex

10% of men and 9% of women report having had anal sex within the past year.

BDSM

14% of men and 11% of women have experimented with sexual sadomasochism, and about 5-10% of American adults has sadomasochistic sex for pleasure at least occasionally. 12% of females and 22% of males have become sexually aroused by a sadomasochistic sex story. 55% of females and 50% of males have become sexually aroused (or had their levels of arousal increased) by being bitten by their partners. If all this still doesn't have you tied up in knots, then consider that 11% of men and 17% of women have tried bondage.

Sex is energy.
Beatrice Wood

Homosexuality

Some gay activists want us to believe that 10% of the population is homosexual. It is true that 90% of both the male and female population considers themselves heterosexual, but this leaves a lot of room for interpretation, as far as the remaining 10% is concerned. Among men and women aged 18-44, 2.3% of men and 1.3% of women consider themselves homosexual, while 2.3% of men and 2.8% of women consider themselves bisexual. Another 3.9% of men and 3.8% of women classify their sexual preference as "other," a number that I confess leaves me wondering, because any given person has only one, the other, or both genders to choose from. It is entirely possi-

Chapter 22
Neither Straight nor Narrow

ble that the "other" category can refer to fetishes and/or *asexuality*, where a partner is not part of the equation.

Still, roughly 7.5% of men and women experience homosexual desire (sexual desire for a person of the same gender), while 6.2% of men and 4.4% of women are attracted to people of the same gender. 2.4% of men and 7.7% of women aged 15-19 report having had sex with another person of the same gender within the past year. Another study indicates that 5% of men and 3% of women consider themselves homosexual.

> *It always seemed to me a bit pointless to disapprove of homosexuality. It's like disapproving of rain.*
> Francis Maude

Again, these numbers don't quite add up to the 10% that some would want us to accept. Still, they are highly significant, because any measurable human trait can range from one extreme (such as total heterosexuality) to the other (total homosexuality), as we learned in Chapter 8. If we assume for just a moment that the latter homosexuality figures of 5% for men and 3% for women are accurate, then the total homosexual population is right about 4% of the overall population. The statistical boundary between homosexuality and bisexuality/heterosexuality occurs right at +/- two standard deviations. Put another way, if we assume that sex with a person of the opposite gender is "normal" (since sex is, after all, about reproduction), and if we then define the range of -2 to +2 standard deviations as "normal," then it becomes easy to predict that 4% of the population will be homosexual. Let me be perfectly clear that I am using the term "normal" in the statistical sense only, without passing any kind of value judgment on homosexuality itself.

> *What do you mean you don't believe in homosexuality? It's not like the Easter Bunny, your belief isn't necessary.*
> Lea DeLaria

As for values, homosexuality is widely frowned upon, and homosexuals have endured terrible persecution over the course of human history, from being strung upside down in frames and sawed in half, to being denied the right to marry. The former is directly attributable to the same folks who brought you the Inquisition, a centuries-long orgy of death and torture wrought at the behest of the Catholic Church. The latter also traces its roots to religion. This begs the question: Is homosexuality aberrant or wrong? If so, then one could reasonably expect humans to be the only species with homosexual members. If not—and if one accepts the hypothesis that humans are just one animal species among many—then one should expect to find homosexual behavior in animals. Bruce

Bahemihl, Ph. D, author of *Biological Exuberance: Animal Homosexuality and Natural Diversity*, found the following:

- 10% of female silver gulls (*laurus novaehollandiae*) are homosexual, and 11% are bisexual.
- 22% of all black-headed gulls (*laurus ridibundus*) are homosexual, and 15% are bisexual.
- 9% of all Japanese macaques (*macaca fuscata*) are homosexual, and 56% are bisexual.
- 100% of all bonobos (*pan paniscus*) are bisexual, and we share 98.5% of our genes with them.
- 44% of all galahs (*eolophus roseicapillus*, a type of cockatoo) are homosexual, and 11% are bisexual.

The roots of homophobia are fear. Fear and more fear.
George Weinberg

If one accepts these numbers, then one must conclude that homosexuality may be the exception rather than the norm (the species must continue, after all), but that there is nothing inherently aberrant or wrong about it whatsoever. I read a number of reviews of Dr. Bahemihl's book. Those who take the greatest issue with his work are extremely likely to resort to the "humans aren't animals" argument that I (among many others) have taken great pains to disprove. These arguments sometimes take the form of "God created animals, but created us in His image." To this I can only reply that all life on Earth shares the same genetic heritage. As we saw in Chapter 4, humans have 26% genetic commonality with yeast (a single-celled organism) and 18% commonality with thale cress (a plant). Seems that God created all life in the same image.

Prostitution

Prostitution and medicine each vie for the title of "world's oldest profession." Think about it: Everyone gets sick at least once in their lives, making an expert in healing a much-needed addition to society. All healthy people have at least some sexual desire, and it therefore stands to reason that some people should be able to make at least a modest living offering sexual services on a fee-based, no-strings-attached basis. The fact is that 69% of white males have been with a prostitute at least once. If we expand the definition of prostitution to cover all paid sexually related services (such as pornographic Web sites, for instance) then that number can only jump dramatically up.

Aren't women prudes if they don't and prostitutes if they do?
Kate Millett

Cheating

> *I always wondered if you clone your wife and have the cloned wife on the moon and the real wife down here, would that be considered cheating?*
> Luis Guzman

Chapter 21 indicated that studies of American and British children revealed that between 5% and 30% of them were not biologically related to their purported fathers. The subject of fidelity has come up several times during our exploration of human sexuality—and rightly so—because it's such a prevalent yet delicate topic. How prevalent? Approximately 20-25% of men and 10-15% of women have at least one affair during their marriage, and infidelity is the #1 cause of cause of divorce in more than 150 cultures.

Who is most likely to cheat? There are many risk factors, some of which include:

- People who have been part of a couple for a long time (especially those who have been together 8 years or longer, since that exceeds the average marriage duration).
- People who have had a higher number of sexual partners prior to marriage.
- Men are more likely to cheat.
- Urbanites are more likely to cheat than their suburban or rural counterparts.
- People who think about sex several times a day are more likely to cheat.
- People who are "pretty happy" in their current relationship are twice as likely to cheat as people who are "very happy."
- People who are "not happy" in their current relationship are four times as likely to cheat as people who are "very happy."

These numbers are low enough to conclude that marriage is a valid institution in many respects. They are also high enough to conclude that "until death do us part" is more than a little unrealistic, especially when viewed against duration and divorce statistics. If marriage itself is valid but the underlying assumptions behind most marriages aren't valid, then the only reasonable course of action is to revise what's expected of marriage so as to make it more compatible with the many biological realities we've discussed throughout this book. I'll present a few ideas in Chapter 23, and will explore this topic in a lot more detail in *The Romantic Savage*.

STDs

We finally come to the downside of all this sex: sexually transmitted diseases, or STDs. One in three sexually active people will contract a sexually transmitted disease by the age of 24, according to a 1998 KFF study. Also in 1998, the National Center for HIV, STD, and TB Prevention (NCHSTP) estimated that at least 65 million people (one in five Americans) are infected with a sexually transmitted disease other than HIV/AIDS.

Have all the sex you want, dear reader. Just please use protection. You probably wouldn't dream of riding in a car sans seat belt or of not wearing a helmet on a motorcycle, where the odds of an accident are far lower than one in five. So why incur the far greater yet just as easily mitigated risk of unprotected sex?

Morality is a venereal disease. Its primary stage is called virtue; its secondary stage, boredom; its tertiary stage, syphilis.
Karl Kraus

Chapter 23

Morality vs. Nature

> *I think one's sexuality can be the center of life, and coming out and discovering your sexuality is something that really can define your existence.*
> Mia Kirshner

We've spent the last five chapters looking at the nature of human sexuality, and how it manifests itself in everything from how each gender views reproduction, to what attracts us to each other, and how pervasive sex is across all societies. It seems only fitting that we conclude by taking a look at our attitudes towards sex to see where they align with—or contradict—our natural urges. Many thousands of pages have been written on this topic; ours will be only the briefest of glances.

Sex is one of the strongest drives a person has because reproduction is life's single over arching priority. Even people who have no intention of having children neither stop nor slow down their sex lives. Well, if sex is the strongest drive a person has, then what better way to exert power and control over that person by regulating her or his sexual activities? And lest anyone think those in power unreasonable, all those in power need do is claim to have the divine authority of some invisible über-alpha male in the sky (God), who is never around to deny anything these people say or do. An instinct as strong as sex needs powerful deterrents if it is to be regulated; hence, many societies devised some extremely cruel and sadistic methods of dealing with sexual impropriety. People (mostly women) are still stoned to death, burned alive, and jailed for violating these sexual codes.

What any god could hope to achieve by creating a complex web of sexual instincts only to seek to curb them is never explained, nor will those in power explain why some individuals (such as King David, the prophet Mohammed, Joseph Smith, and others) are able to openly commit sexual acts that would get any lesser person tortured and/or killed. Muslim women can face extreme penalties from flogging to death for having premarital sex, and yet young (and sexually repressed) men are encouraged to blow themselves up in the name of Allah in order to receive 72 virgins in the afterlife, a clear violation of both the law allowing up to four wives and restrictions against premarital sex.

Both the Bible and the Koran restrict the sex drive to an extraordinary degree, only to specifically allow men to have their way with women captured during wartime, the women's marital status notwithstanding. This incentive has proven plenty powerful; untold millions of young men have marched off to war after holy war, many of them presumably seeking nothing more than the pleasure of female company forbidden to them at home. I have to wonder whether 68 of the aforementioned 72 virgins get punished for having sex with the deceased martyr, or whether they fall under the "captured in warfare" exclusion.

> *The Christian churches were offered two things: the spirit of Jesus and the idiotic morality of Paul, and they rejected the higher inspiration. Following Paul, we have turned the goodness of love into a fiend and degraded the crowning impulse of our being into a capital sin.*
> Frank Harris

Do all of these restrictions work? As we've seen in the previous five chapters, the answer is a loud, unequivocal no. No culture can stop people from having sex. Life always finds a way, no matter how those seeking power try to rig the game.

Monogamy

Most modern nations only sanction marriage between one man and one woman. Exceptions (such as Mormons) are generally frowned upon. Also, a growing number of modern nations are allowing homosexual marriages, which does nothing for reproduction, but everything to guarantee this fundamental human right to all people in love. By contrast, primitive tribes tend to be more polygamous, with 85% of these societies allowing men to marry more than one woman. Polyandry (one woman with multiple husbands) is virtually unheard of, with the Fore tribe being a notable exception (and this primarily for allocating land). This wide flexibility shows that cultural

> *I would say that the surest measure of a man's or a woman's maturity is the harmony, style, joy, and dignity he creates in his marriage, and the pleasure and inspiration he provides for his spouse.*
> Benjamin Spock

standards are the single largest influence on available marriage options.

All cultures experience some degree of extramarital sex, a testament to the evolutionary novelty of pure monogamy. Here again, attitudes and laws vary widely, with some cultures still adhering to the "adulterers must die" laws found in the Bible and Koran, and others merely making adultery grounds for divorce. How one is raised also plays a huge role in individual attitudes towards extramarital sex. The net effect of marriage, adultery, and divorce is that humans are somewhat polygynous in even the most nominally monogamous societies. Males, not surprisingly, are usually the ones who initiate changes in sexual relationships.

Adultery & Divorce

Ah, yes, divorce... from the Latin word meaning to rip out a man's genitals through his wallet.
Robin Williams

A close reading of adultery laws reveals that their primary purpose is to guarantee paternal confidence, that is, that a child being raised by its father really is the father's biological offspring. Remember that cuckoldry represents the worst possible loss from an evolutionary standpoint. Hence, adultery is most often defined based on the woman's actions, with the man's actions being far less important. For example, the Bible holds that both a divorced woman and her lover are adulterers. Of course, there is no restriction on men remarrying.

Sex & Violence

You never need an argument against the use of violence, you need an argument for it.
Noam Chomsky

If sex is one of the strongest instincts we have, then it follows that sex lies at the root of much of the violence that occurs in the world. I've pointed out the inherent conflict between men and women in Chapter 14, and in the previous five chapters. This conflict revolves around paternal confidence and extends to children. Stepchildren are far more likely to be abused than biological children.

A study of 42 murders caused by sexual jealousy in Detroit in 1972 found that 33 of these cases were men killing either the woman (16 cases) or the other man (17 cases). The remaining 11 cases had the man killed by either the woman (9 cases) or her relatives (2 cases). The men in this study were equally likely to kill the woman or rival male. Women who killed were twice as likely to kill the man instead of the rival woman.

Sexual conflict between men and woman seems to revolve around issues of paternal confidence, lack of which also spawns conflict among children. Stepchildren are far more likely to be abused than biological children. The most derogatory terms one can use against a woman involve her reproductive functions, be they her actual reproductive organs (*cunt* is both slang for the vagina and the single worst insult one can use against a woman) or her sexuality (terms like *whore*, *slut*, *tramp*, *floozy*, etc. fall into this category). Call a woman any one of these names, and she'll fight back with everything she has. I can't blame her one bit. Men who have many sexual partners are generally regarded as role models, while promiscuous women are denigrated. This reflects the obsession with paternal confidence brought about by the inherently unequal investments in raising children.

Religion

Let's take a closer look at religious views of sex. As you read, compare what you see against what you know of modern sex laws, and you'll see the extent to which religion has influenced modern rules and customs. Be also on the lookout for acts that modern laws prohibit, despite being explicitly permitted in the religious texts themselves. And finally, given what we've learned so far about both human nature and the nature of human sexuality, the extent to which religion tries to exert control over the general population by restricting its most fundamental drives should become extremely self-evident.

All religion, my friend, is simply evolved out of fraud, fear, greed, imagination, and poetry.
Edgar Allan Poe

The Bible

The Bible is laden with rules regarding sexual conduct and penalties for sex offenses. It is also full of stories of people who openly flout the rules, and either get away with it, and/or receive high praise for their righteousness. Here is just a small sample of what the Bible has to say about sex:

Religion is the masterpiece of the art of animal training, for it trains people as to how they shall think.
Arthur Schopenhauer

- The book of Leviticus contains rules against incest, and specifically defines the relatives with whom sexual relations constitutes incest.

Christian teaching about sex is not a set of isolated prohibitions; it is an integral part of what the Bible has to say about living in such a way that our lives communicate the character of God.
Rowan D. Williams

I know of no book which has been a source of brutality and sadistic conduct, both public and private, that can compare with the Bible.
James Paget

If one were to take the bible seriously one would go mad. But to take the bible seriously, one must be already mad.
Aleister Crowley

- Homosexuality is "disgusting" and is prohibited, although, interestingly enough, the passage (Leviticus 18:22) applies to men having sex with men, with no mention of lesbian sex.

- The Old Testament speaks of women unable to bear children. Infertile men are never mentioned. Apropos of men, there seems to be a disproportionate percentage of male children throughout the Bible.

- Every man must be circumcised. King David paid a dowry of 200 foreskins for his wife, twice the asked price of 100 foreskins. Why anyone would see the severed tips of penises as currency, or what use one could put them to, is not explained.

- Lot offered his virgin daughters to be raped and used by men who were out to rape male angels. This same Lot has sex with his daughters (of course they get him rip-roaring drunk and have their way with him), who bear sons. The Bible praises Lot for his righteousness despite the fact that the sexual laws spelled out in the book of Leviticus explicitly prohibit his conduct.

- Abraham made a servant hold his testicles while swearing an oath to God, this despite plenty of admonitions against impurity and homosexuality. Abraham is not alone in performing this bizarre ritual.

- Modern Christian rules against masturbation and birth control seem to be inspired by Genesis 38:8-10, where Onan masturbates instead of having sex with his dead brother's wife.

- Menstruating women are unclean and must be kept away from all others until they are "clean" once more.

- Captured virgin women can be used at will.

- A man who doesn't like the woman he marries can claim she wasn't a virgin when she married. If the father can't produce the bloody bedsheets to contradict the claim, the woman must be stoned to death on her father's doorstep.

- Adultery is punishable by death.

- Women who don't scream loudly enough while being raped must be stoned to death.

- King David got away with adultery because the woman had been somehow purified by means unknown.

- Bestiality is prohibited as unnatural.

- Prostitution is a deadly trap that causes infidelity.

- Husbands and wives should not deny each other sexual intimacy.

- Run from anything that stimulates youthful lust (and) follow anything that makes you want to do right (2 Timothy 22). How, exactly, teens are to control their evolutionarily-programmed hormones that are driving them to seek mating opportunities at all costs, is conveniently not mentioned.

- Christ took the definition of adultery one step further by pronouncing that anyone who so much as looks at a women with lust in his eye has already committed adultery. An eye that causes one to feel lust should be gouged out and thrown away. Ouch.

- As for divorce, Christ said that a man who divorces his (faithful) wife causes her (and anyone who marries her) to commit adultery. Divorcing an unfaithful wife seems to be OK, however. Also, if adultery is defined as sex with a married person, how can having sex with a divorced (unmarried) person be called adultery?

- Christians are supposed to control their bodies, eschewing "pagan" lust (but the fact that many Christian rituals were adopted—if not outright copied—from pagan rituals is, naturally, perfectly fine).

- There are several accounts of God "visiting" women (Sarah and Hannah come to mind) and giving them sons. Christ therefore cannot be "God's only begotten son," unless I missed something.

- And, lest we forget, women are guilty of original sin, thanks to Eve who ate the forbidden apple.

It should be plainly obvious that the Bible contains numerous rules that fly in the face of everything we know about evolu-

---

It ain't those parts of the Bible that I can't understand that bother me, it is the parts that I do understand.
Mark Twain

The Bible and the Church have been the greatest stumbling blocks in the way of women's emancipation.
Elizabeth Cady Stanton

tion and human sexuality. The Bible also freely allows double standards, where some of God's children are clearly more equal than others. I haven't even touched the many inconsistencies and flat out contradictions that exist in the Bible. More in *The Divine Savage*.

The Koran

If my children do not behave according to Islam, if they do not pray for instance, I will punish them.
Abu Bakar Bashir

Jews and Christians have no monopoly on sexual repression. Here are a few choice edicts from the Koran regarding men, women, and sexuality:

- Men may not marry non-Muslim women. A Muslim slave is deemed superior to a nonbeliever free woman.

- Menstruating women are filthy and polluted.

- Women are sexual objects for the husband's taking whenever he chooses.

- Women have fewer rights and lower legal status than men.

- Men are allowed to marry up to four women, provided they all receive equal treatment, and may capture women during war and use them however they like.

- Men who are unable to marry women may marry girls.

- Men must abstain from sex unless married, except that they can have their way with captured women. Married slave women are fair game.

- Mohammed was, of course, exempt from the rules limiting the number and treatment of wives. None of his wives could venture outside the house. He married his adopted son's wife and was allowed to divorce his wives any time he wanted, to trade them in on more submissive replacements. No one could marry his widows after his death.

It seems to me that Islam and Christianity and Judaism all have the same god, and he's telling them all different things.
Billy Connolly

- Lewd women should be incarcerated for life.

- Husbands should beat wives who refuse to obey them.

- Adulterers and fornicators should receive 100 lashes (except men who capture and rape women during wartime, of course).

- Women are required to wear veils and cover their bodies when outside their homes.
- Divorced women must receive support from their husbands until they can support themselves.

Mormonism

Mormonism is a newcomer among the world's religions, with the *Book of Mormon* having been published in 1830. It seems that many of the rules and guidance about sexual matters come from teachings by various church officials. Here is just a small sampling.

- Men are allowed to have adulterous sex nine times. It is only on the 10th time that they finally become guilty.
- A Mr. Benjamin Covey was excommunicated for having sex with his underage (less than 12 years old) foster daughters, but was later readmitted to the church and allowed to serve as a bishop.
- Plural marriage (polygamy) is expressly allowed.
- There exist several communities of "Mormon fundamentalists," where men have numerous wives and routinely marry underage girls and commit various forms of incest. Jon Krakauer, author of *Under the Banner of Heaven*, describes a woman showing him a family tree that he said resembled an intricate engineering diagram.
- Husbands were encouraged to abstain from sex for seven days following their wive's menstrual cycles so as to maximize the odds of successful conception.
- Women who bear sons should be separated as unclean for 40 days, while women who bear daughters should be separated for 70 days.
- First Counselor Herber Kimball recommended decapitating adulterers.
- Women married to infertile husbands may seek proxies in order to bear children (polyandry).
- Oral sex is prohibited as being unnatural, impure, and/or unholy.

There is all the difference in the world between teaching children about religion and handing them over to be taught by the religious.
Polly Toynbee

At some point we must make a decision not to allow the mere threat of charges of cultural or religious insensitivity to stop us from dealing with this evil.
Armstrong Williams

- Married couples must cover their nakedness (by wearing special undergarments), and can't so much as talk dirty in bed. Imagine not being able to see your spouse naked!

- Masturbation is a big no-no.

- Homosexuality is another big no-no.

- Sexual sins are tantamount to murder (though plenty of Mormons have gotten away with murders committed in the interest of sexual propriety).

- Tongue kissing (so-called *French kissing*) and petting are considered fornication.

- Virtue is prized above life. True Mormons are said to prefer burying their dead children over having them lose their chastity. Virtue and chastity are synonymous.

- Like Judaism, Christianity, and Islam, Mormons too have their share of bigwigs who got away with sexual acts that would condemn most of lesser status.

> *I hope I never get so old I get religious.*
> Ingmar Bergman

Buddhism

Buddhists have a far more pragmatic approach to sexuality. Here is an extremely brief summary:

- Buddhist monks are not allowed to have sex, as this can interfere with the path to enlightenment.

- In general, men need not show much restraint, except for commonsense restrictions against sex with girls, married women (adultery), convicts, or women engaged to be married.

- There is nothing inherently wrong or bad about sex. Sexual pleasure is fine; attachment (addiction) to sexual pleasure is a problem.

> *The mind is the Buddha, and the Buddha is the mind.*
> Bodhidharma

Hinduism

The Hindus gave us tantra and the Kama Sutra. Enough said.

> *Hinduism's basic tenet is that many roads exist by which men have pursued and still pursue their quest for the truth and that none has universal validity.*
> Kenneth Scott Latourette

Secular Law

There are many thousands of laws designed to regulate sexuality on the books across each of the United States and the entire world. Many of these laws specify such things as:

- Minimum age of consent (below which the crime of statutory rape is deemed to have occurred).

- Definitions and penalties for incest.

- Definitions and penalties for sodomy (which, strictly speaking, refers to anal sex, but which has been defined as meaning any "unnatural:" sex act). Examples include prohibitions against oral sex, cohabitation (unmarried people living together, sometimes even for just one night), etc. Thankfully, a United States Supreme Court decision in 2003 rendered state sodomy laws unenforceable.

- Definitions and penalties for forcible rape.

- Definitions and penalties for incest.

If one commits the act of sodomy with a cow, an ewe, or a camel, their urine and their excrements become impure, and even their milk may no longer be consumed. The animal must then be killed and as quickly as possible and burned.
Ayatollah Khomeini

Lessons Learned

Monkey see, monkey do. Humans learn a tremendous amount about life and how to live it by observing and mimicking those around them, a fact that helps preserve the rich variety of cultures found on Earth. It should not be surprising that children take their cues about appropriate sexual conduct from their parents and society at large. For example, children of divorced parents marry younger, have children sooner, and are more impulsive and aggressive than children of married parents. I suspect that much of this has to do with the nastiness that accompanies far too many divorces, a sad by-product of the religiously-based expectation that marriage should be a lifelong thing (in direct violation of everything we know about human sexuality). Girls in fatherless households hit puberty 6 months sooner than girls whose fathers are present.

The evolutionary reasons for this behavior are obvious: If one expects a relationship to be short-term, then one needs to get about the business of establishing a genetic legacy sooner. Those who expect long-term relationships benefit by waiting to make sure they have found the best possible situation before reproducing. It's a variation of the "ghetto reproduction" strategy we discussed in Chapter 20.

Homosexuality

Homosexuality in Russia is a crime and the punishment is seven years in prison, locked up with the other men. There is a three-year waiting list.
Yakov Smirnoff

Attitudes towards homosexuality vary widely. Some cultures accept and even embrace it as part of the broad range of human sexual potential, while others actively repress it as evil. The United States is home to groups seeking to extend marriage rights to homosexuals, while others seek to pass Constitutional amendments defining marriage as occurring between one man and one woman (the so-called "Defense of Marriage Act"). The justifications proffered for prohibiting homosexual marriage are all religiously based. It is interesting to note that most (if not all) religious texts define homosexuality as occurring between two men, while not even mentioning lesbianism. I can only wonder whether this is a result of patriarchal societies and the seemingly innate fascination and curiosity that most—if not all—heterosexual men have about lesbianism.

The simple truth is that homosexuality is perfectly natural, a probable by-product of some genetic nuance that can occur just like any other genetic nuance. So-called therapy designed to turn gay people straight does nothing but confuse and depress the recipients of this therapy. If you don't like homosexuality, then don't have sex with someone of your own gender, and leave it at that. My friend Irv recounted the following conversation that a child had with his mother (who is a friend of Irv's):

"Mother," said the child, "I'm gay."

"So?" replied the mother, glancing up from her magazine, "Do I need to bake or knit anything?"

Should Logan ever tell me he's gay, my response will be a simple, "Yes, and...?"

296 | *The Natural Savage*
Discovering the Human Animal

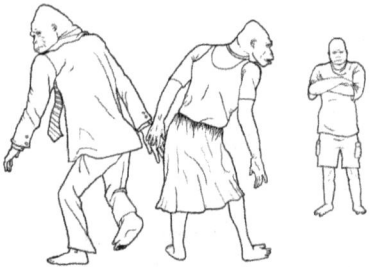

PART SIX

The Great Unknowns

298 | *The Natural Savage*
Discovering the Human Animal

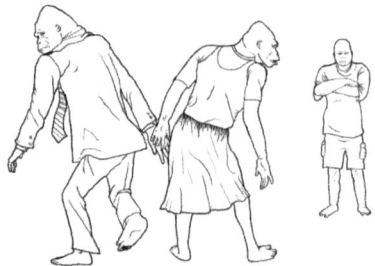

Chapter 24

Death: The Final Frontier?

Even death is not to be feared by one who has lived wisely.
Buddha

You are going to die.

You may not die today, tomorrow, or for many years, but die you will, whether you like it or not. That is your inevitable destiny, one shared by all living things on Earth, because death is the final and most permanent of the six core life functions introduced in Chapter 2 and discussed throughout this book. Nothing is forever, at least not in the physical plane of existence as we know it. Furthermore, you owe your very existence to countless plants and animals that have died to nourish you. By the time you leave this world, you will have consumed between 60,000 and 80,000 pounds of food (that's one or two full truckloads), all of which came from plants and animals. Life requires death, both for lebensraum (the planet would run out of room within days sans death) and nutrition.

I look upon death to be as necessary to our constitution as sleep. We shall rise refreshed in the morning.
Benjamin Franklin

Place yourself into the grand scheme of things, and you'll soon realize that future generations are relying on your death in order to live. Since life exists to beget new life, your dying is quite literally the single greatest service you can do for your descendants after giving birth to them. The countless molecules that comprise your body will move into, nourish, and become part of new organisms, thus guaranteeing you at least a modicum of immortality, in the same way that your body consists entirely of molecules gleaned from what you've eaten.

Why is this true, and how does the mechanism of death work? While we're on the topic, does the death of our physical bodies and the brains from which consciousness presumably originates mean that this lifetime is the only one we get? Let's look at these questions.

Life Beyond Reproduction

As if human sexuality and reproduction weren't aberrant enough by animal standards, it turns out that humans (specifically human women) are the only animals on Earth who experience menopause, the cessation of fertility that comes years, or even decades, before death. At first glance, this seems to go against evolution's single-minded goal of passing on one's genes to the next generation.

Further examination resolves this problem. The neoteny we discussed in Chapter 8 means that humans are born helpless, and remain helpless for many years, placing a huge burden on their parents. The complexities of life also require many years to instill, especially when one includes rapidly changing technology as part of a child's necessary education. Add in the male's competing reproductive interests, and suddenly having an extra set of hands around the house to help becomes more than just a luxury. A woman who helps her daughter raise grandchildren helps ensure the continuance of her genetic legacy, especially since the grandmother has already made the huge investment in her daughter. Remember that there is nothing a woman can do to increase her lifetime reproductive potential beyond the 10 to 30 child maximum, leaving her little choice but to throw all she has into every single child. Men can increase their reproductive potential as easily as having sex with more women. Men also never completely lose fertility.

Menopause is an evolutionary oddity that addresses the unique requirements of human reproduction. As ironic as it sounds, the fact that women cease being fertile long before they die allows more human children to reach adulthood and go on to have children of their own.

Rock and menopause do not mix. It is not good, it sucks and every day I fight it to the death, or, at the very least, not let it take me over.
Stevie Nicks

Just look at my face. Its an extraordinary experience. All of my friends who are grandparents have been saying, just wait, a bit cynically, but its just extraordinary. You feel like a child again yourself. Just walking on air.
Blythe Danner

Repair or Recycle?

> *How pleasant it is for a father to sit at his child's board. It is like an aged man reclining under the shadow of an oak which he has planted.*
> Voltaire

An airplane designed to haul heavy loads a long distance (such as a jetliner) trades speed and maneuverability for efficiency. A fighter trades efficiency for speed and maneuverability. Every airplane design represents a compromise that makes the plane better at its chosen mission, at the expense of other potential capabilities. Airplane design is a set of trade-offs. Evolution is another set of trade-offs. An animal built for speed (such as a cheetah) can't afford to haul large quantities of fat with it, meaning that speed comes at the expense of being able to go without food.

Every living thing runs the risk of injury and/or illness. The question is what do to about that risk. A body can only metabolize so many calories in a given day. Does it make more sense to allocate those calories to self repair (healing), routine maintenance (replacing cells), or reproduction (mating/pregnancy/birth/etc.)? Put another way, how should any given species allocate its calories in order to be able to produce as many children as possible?

The average human male requires over 1,600 calories per day just for routine maintenance, the average women over 1,400. That's a lot of food that, as I mentioned above, translates into 30-40 tons in the average lifetime. Nature invests a lot in each human. Moreover, most humans face relatively little risk of sudden death unlike, say, a mosquito that can be snapped up in midair without any warning whatsoever. And, lest we forget, human children require a lot of help in order to survive, much less flourish. All of these factors make it essential to invest in ways to repair the damage that will almost inevitably occur during a lifetime. The human immune system and ability to heal from most wounds is Nature's insurance policy.

> *Just like those who are incurably ill, the aged know everything about their dying except exactly when.*
> Philip Roth

An insurance adjuster's job is to assess damage and make the decision whether the damaged item (such as a car) can be repaired, or whether it's "totaled" and should be scrapped. Evolution acts as an insurance adjuster by determining what injuries we can recover from and what we can't. Most injuries can heal; however, a destroyed or removed organ or limb is gone forever. It's one thing to close a cut or fuse a broken bone since skin, blood, and bone cells are being created all the time. Regrowing an entire limb that required an entire lifetime to get to its pre-amputation state requires far too much energy and is therefore a waste of resources—not least because a person with a missing limb is at a distinct disadvantage for escaping predators, avoiding infection, finding food, etc. If the cost of repairing a given injury is less than the reproductive benefit of keeping you around, chances are your body can and will repair itself from that injury. If not, then chances are your body cannot and will not repair itself.

Remember that evolution is a numbers game that is every bit as dispassionate as the triage performed at the scene of a major accident with multiple casualties. Whereas one critically injured person might be recoverable if s/he is the only victim, stopping to help her or him when there are dozens or hundreds of victims could mean sacrificing others who require a lot less effort to save (such as simply placing direct pressure on a bleeding wound, which takes only seconds to do, as opposed to time-consuming treatments such as inserting a breathing tube and supplying artificial ventilation/respiration). Evolution doesn't care about you, the individual, in the slightest. Evolution occurs on behalf of the entire species.

Two aged men, that had been foes for life, met by a grave, and wept—and in those tears they washed away the memory of their strife; then wept again the loss of all those years.
Jean Paul

Universal Deterioration

Say you're designing a machine with a planned useful life of, say, 70 years. It does you little good to build the machine out of cheap junk that will require constant fixing in order to keep it running for the full 70 years. It also does you little good to include parts that can last longer than the design life. Your goal should be to design a machine where all of the parts will wear out together. Individual circumstances will require repairs to each machine, and individual components that go the distance on one machine may require replacement on another machine; that is the way of things. Still, your overall goal should be to design a machine that will have as few useful parts left as possible when it hits the 70-year mark. Otherwise, you're spending too much to build each machine, an error that will add up over time with multiple copies of the machine being built.

Same with natural selection. A body is a machine with a certain life expectancy. Modern medicine can repair or replace individual components in order to prolong the inevitable, but

A dying man needs to die, as a sleepy man needs to sleep, and there comes a time when it is wrong, as well as useless, to resist.
Stewart Alsop

For a person who is dying only eternity counts.
Friedrich Durrenmatt

> *I am afraid of death, scared by it. I already don't know whether I exist or not. So dying really terrifies me.*
>
> Stephen Rea

nothing can stop the overall aging process, which represents a gradual collapse of all bodily systems at once. This collapse continues until everything shuts off at once (so-called "death by natural causes"), or until one critical component gives out and takes the rest of the body with it (such as a myocardial infarction or "heart attack"). Illness and injury are Nature's trump cards, always waiting to prey upon those unlucky enough to be in the wrong place at the wrong time. I must point out that many (if not most) cases of individual components failing are caused by our modern lifestyle (such as foods that bear little resemblance to our primordial diets).

I'll say this yet again: The sole evolutionary purpose for living is to create new generations of life. Evolution therefore favors strategies that maximize reproductive potential. Every creature's life expectancy is perfectly tailored to that species's unique abilities and limitations, and humans are no different. We live long enough to have children, raise them, pass on our incredibly rich and complex knowledge to the younger generations, and that's it. Once we have accomplished that mission, nature has no more use for us, and we must move aside (die) to leave room and resources for our offspring and their offspring. A body designed to have individual components wear out while the body itself lives on consumes resources that could go towards young, healthy individuals. Worse, the partially worn out body may have to depend on others for feeding, cleaning, etc. This is a total waste of resources from an evolutionary standpoint, and indeed it is only very recently that humans have developed the technology, abundance of resources, and professions dedicated to doing just that. Thus, nature has designed bodies to undergo an overall collapse that renders attempts at component-based repair impossible.

> *I don't believe in dying. It's been done. I'm working on a new exit. Besides, I can't die now—I'm booked.*
>
> George Burns

Life without Death

> *I'd rather be dead than dying.*
> Stephen Evans

What if death did not exist? It can be tempting to think of life without end, a life where you would get to meet your distant descendants hundreds, thousands, or even more years into the future. But could such a scenario work?

No.

A world without death must by definition be a world without life. Whatever species first evolved would expand to cover the entire planet, and then run out of room to keep on evolving. This species would have nothing to eat, unless it consumed the rocks and water on which it lived. This could potentially work in a two-species environment, where one eats rock and water while the other consumes the waste from the first species and turns it back into rock and water (just as trees consume our "waste" carbon dioxide and release oxygen for us to breathe). But alas, rock doesn't form in an instant. And how could these species possibly survive the plethora of natural disasters from earthquakes to asteroids that have eliminated so many individuals and species over the eons?

Furthermore, you and I would not be here to have this discussion. A planet without death could only sustain so many individuals. Without death, there would be no future generations, no evolution, no you. If death were to stop today, the entire planet would be several feet deep in insects within days—not a pleasant thought. Death, mine and yours included, thus increases the chances of our winning the evolutionary game by leaving behind our genetic legacy. If you have aging parents, chances are great that they've repeatedly told you that they don't want to be a burden to you and your family. They realize, either consciously or unconsciously, that their job is done, and that their days are numbered. This is why the suicide rate is so high for old people. The National Institute of Mental Health reports that suicide rates among people 65 and older are 1.3 times higher than the average national suicide rate. The suicide rate among white men aged 85 or older is over 1.6 times higher than the national average.

Life without death would be impossible. That's the bad news. The good news is that you are leaving behind nutrition for a wide array of plant and animal species, who will in turn provide nourishment to future generations of people. At least that's the way it used to work, before modern burial and embalming techniques prevented us from truly returning to nature, and thus, ironically, denying us that most fundamental level of immortality.

I say to people who care for people who are dying, if you really love that person and want to help them, be with them when their end comes close. Sit with them—you don't even have to talk. You don't have to do anything but really be there with them.
Elizabeth Kubler-Ross

If even dying is to be made a social function, then, grant me the favor of sneaking out on tiptoe without disturbing the party.
Dag Hammarskjold

Design or Accident?

> *Fix reason firmly in her seat, and call to her tribunal every fact, every opinion. Question with boldness even the existence of a God; because, if there be one, he must more approve of the homage of reason, than that of blindfolded fear.*
> Thomas Jefferson

I believe it impossible to discuss death without at least touching on some of the deepest questions ever asked: Is this all there is? Did the universe begin through quantum fluctuations or other "natural" causes? If so, what caused those causes? Did life evolve thanks to a series of random events? Or is this all the result of some deliberate effort (creation)?

As far as I am concerned, evolution is a proven fact, to the extent that anyone can truly prove anything. My belief in evolution is as fundamental as my belief that 2+2=4. Modern genetic research shows that humans are still evolving, and quite rapidly at that. Contrary to popular belief, we are not evolving through interbreeding into a single pan-racial ethnicity; rather, we are evolving into entirely separate species. If this continues, then people will eventually be unable to freely reproduce with any biologically mature (and non-menopausal) adult of the opposite gender. I can only wonder how concepts such as human rights, the equality of all persons, etc. will stand up to that brave new world.

> *That deep emotional conviction of the presence of a superior reasoning power, which is revealed in the incomprehensible universe, forms my idea of God.*
> Albert Einstein

Does belief in evolution rule out the possibility of creation and of a creator (god)? Does evolution have to be at odds with the ideas of soul, afterlife, reincarnation, etc.? There are plenty of people, from priests on one extreme to Richard Dawkins on the other, who trumpet an either-or answer to this question. Why? Why couldn't a god or other conscious creative force have created evolution? Believe it or not, substituting some very simple metaphors and making allowances for a primitive audience that the writer is trying to impress, reveals that Genesis 1 tells the scientifically accepted history of the universe perfectly, right down to the exact order in which matter formed and life evolved. This leaves open the question of whether (mostly Western) scientists knowingly or unknowingly gerrymandered observations to fit into the Biblical model or whether the Bible itself is on to something. I delve into this topic in much more detail in *The Divine Savage*. Meanwhile, suffice it to say that there is no inherent conflict between the ideas of creation and evolution beyond the perpetual pissing contests taking place between extremists on both sides of the question.

Self-Awareness

Life in general has no inherent need for consciousness. Humans, with our extremely complex lifestyles (especially where reproduction is concerned), benefit from self-awareness, and from an awareness of others that allows them to, among other things, forge tight social contracts for mutual security by forgoing harmful acts. Our primordial means of locomotion by climbing and swinging through trees also benefits from the ability to see one's self as a separate entity moving within a larger environment, as noted in Chapter 3. The fact that we think of ourselves as unique individuals is a very novel thing among animals, most of whom can't pass the "red dot" test, where researchers paint a red dot on a sleeping animal's forehead to see if it wipes it off when it wakes up and peers into a mirror. Still, our having such highly developed consciousness is not of itself an argument for design, creation, or soul.

Without consciousness and intelligence, the universe would lack meaning.
Clifford D. Simak

No Waste

Nature wastes absolutely nothing, and is the only "machine" that operates with 100% efficiency. Every measurable trait of every plant and animal species exists for the benefit of both the individual organism and the entire species. Science used to think of much of the DNA carried by plants and animals (humans included) as evolutionary junk, in much the same way the appendix was thought to be a useless vestigial organ. Recent research has discovered that this so-called "junk" DNA is not junk at all. It is now apparent that every single piece of our genetic blueprint pulls its own weight by having a valid and essential function. And the appendix? Turns out that it plays an important role in building a child's immune system. I certainly don't know of anyone who seriously suggests than an immune system is useless junk!

A man who dares to waste one hour of time has not discovered the value of life.
Charles Darwin

If there is absolutely no waste in nature, then our mere ability to ask and ponder life's big questions is either the one exception that proves the rule, or it serves a valid purpose. As the saying goes, the race is not always to the swift nor the battle to the strong, but that's the way to bet. Believing that our ability to ponder these questions (which are absolutely useless for daily life as we know it) serves some larger purpose is therefore far more logical than believing that we've found the one example of Nature's wastefulness among endless trillions of possible examples. I therefore believe that our drive to answer these questions must serve some larger purpose. My entire *Savage* series of books is the realization of my own quest for answers, which I am undertaking primarily to find my own answers—a far higher purpose than simply trying to sell some books. If my quest helps you find your own answers or ask more questions, then so much the better.

The Ascension Fallacy

People seeking to discredit creation point to the fact that life has seemingly evolved in an upward direction, from the most primitive single-celled organisms, to the most complex creatures, including humankind. This argument has a number of problems, including:

- *Irreducible complexity.* Remove one or more components from structures as diverse as eyes and cilia (small hairs inside our bodies that do things like move food through out intestines, eject mucus from our lungs, propel sperm, etc.), and they simply fail to work. It is difficult if not impossible to explain how any species lugging around a useless "work in progress" could hope to survive in a world filled with competitors lacking such an impediment. At least that's what some folks think.

- *Evolutionary algorithm.* Richard Dawkins and others assert that evolution can take place in stages, with the desired end product specified by an "evolutionary algorithm." Thanks to such algorithms, the process of evolution can occur rapidly with as few intermediate steps as possible. This argument fails in at least two places: First, isn't "algorithm" another word for "blueprint," and doesn't that signify design, the very thing Dawkins is trying to disprove? Second, why keep evolving once the algorithm has fulfilled its purpose, and why and how does the algorithm keep changing? Remember that *homo sapiens* is evolving into separate species as I type this very sentence and as you are reading it.

- *Randomness.* Imagine yourself suddenly teleported to a completely random city you've never visited or even

---

The day which we fear as our last is but the birthday of eternity.
Lucius Annaeus Seneca

An expert is a person who avoids the small errors while sweeping on to the grand fallacy.
Steven Weinberg

Biology is the science. Evolution is the concept that makes biology unique.
Jared Diamond

heard of before. Now imagine yourself trying to navigate this strange place. You would invariably find yourself going down dead-end streets, possibly going in circles, and probably some combination of the two. There is no reason to suspect that you'd move in a deliberate direction without having to backtrack, nor is there any reason to think that the process of evolution plays by different rules. This does not mean that your actions are random; each turn you make represents your best attempt at reaching your goal, given the circumstances at the time you made your decision. In fact, there is no such thing as "random," only those events for which we lack complete data and/or for which we'd rather not do the math. Even our inability to predict quantum phenomena beyond a fraction of total certainty does not randomness make. Still, one does not need randomness to exist in order to make a long series of wrong turns. Again, there is no reason to think that evolution does not follow this general rule.

- *Religion.* I discuss religion in far more depth in *The Divine Savage*. For now, research indicates that religion seems to be literally programmed into our genes, as does our enduring belief in a creator and of life beyond this one. It is easy to dismiss this as wishful thinking designed to ease the sheer terror of contemplating our inevitable death, and/or as the logical extension of our hierarchical nature that enables efficient group living and its associated predator protection to exist. Please note that I am using the term *religion* merely as a label for these beliefs, not as any reference to any specific organization/church/etc.

- *Sameness of life.* All life shares a common genetic legacy, as I've pointed out in several places throughout this book. One need look no further than bone patterns across fish, birds, reptiles, and mammals to see the striking similarities. Auto makers routinely borrow parts from one car to make other models, which can each have very different functions. It is not uncommon for a sports car to share many of its parts with an SUV, for example. And why not? If one has a successful design, then it's wasteful to change for change's sake.

Unless one is a religious fundamentalist and believes that man was created in the image and likeness of God, it is foolish to believe that human beings are exempt from biological classification and the laws of evolution that apply to all other life forms
J. Philippe Rushton

We are so arrogant, we forget that we are not the reason for evolution, we are not the point of evolution. We are part of evolution. Unfortunately, we believe that we've been created to dominate the planet, to dominate nature. Ain't true.
Ted Danson

> *Theology made no provision for evolution. The biblical authors had missed the most important revelation of all! Could it be that they were not really privy to the thoughts of God?*
> E. O. Wilson

- *Inherent design.* The simplest living thing makes the most complex human-made contraption pale by comparison. The cell will invariably be more complex and infinitely more efficient than even the very best human inventions. We cannot look at so much as a chipped rock without ascribing design to it, so how can we look at the marvel of life as anything less? As I alluded to above, simply tossing a pile of parts into the air in hopes they'll land in the form of a working pocket watch is ludicrous, as is expecting a tornado tearing through a junk yard to result in a totally airworthy airplane.

- *Inherent optimization.* Adding, removing, or changing parts on a whim very rarely improves the device being tampered with. In fact, doing so most often risks harming or perhaps even destroying said device. Why should evolution be the exemption?

- *Single-mindedness.* Evolution doesn't exist to make us happy; we exist to make evolution happy by passing on our genes. Interestingly enough, most people are the happiest when doing things that bolster their evolutionary success. In other words, people are the happiest when they are doing things that enhance their ability to carry out the six core functions of life described in this book. This is a clear sign of design. In fact, among human artifacts, the more specific the purpose, the more extensive the design effort required to pull it off.

> *The argument for intelligent design basically depends on saying, 'You haven't answered every question with evolution.' Well, guess what? Science can't answer every question.*
> Kenneth Miller

I could go on and on, and do in *The Divine Savage*. For now, suffice it to say that I believe that the "design-less" theory of evolution has about as many holes as the ancient *terracentric* (Earth-centric) models of the solar system that required layers upon layers of epicycles (orbits within orbits) to describe the motions of the heavens as seen from Earth. In this example, our having the humility to remove ourselves from the center of everything and to see beyond a "perfect" circle yielded the simple marvel of a solar system whose planets move in regular elliptical orbits, and where the fuel source for all life on this planet, the Sun, has its rightful place at the center of our little corner of the cosmos. Occam's Razor calls upon us to accept the simplest possible explanation when presented with multiple possible answers. Is it easier to believe in a long series of freak occurrences, or in a conscious creative force? You decide.

How Little We Know

There are two separate things happening inside our heads. For this discussion, *mind* represents everything we are consciously aware of. *Brain* represents both everything we are unaware of, and the physical organ that contains it all. It is tempting to believe that mind rules brain according to the theory of "mind over matter." The truth, as we discovered in Chapters 4 and 5, is that our minds have very little idea about the vast majority of what's going on in our brains.

We allow our ignorance to prevail upon us and make us think we can survive alone, alone in patches, alone in groups, alone in races, even alone in genders.
Maya Angelou

What exactly is consciousness? Are our brains the source of consciousness, or merely transceivers for consciousness? Proponents of the former argue that we have no direct perception of any existence beyond this one, and that such an existence therefore cannot be. To this, I reply that a scuba diver cannot directly perceive land and sky, but can at least make strong inferences about their existence from what s/he can observe (such as a rising bottom that clearly extends above the waterline, alternating light and darkness above the waterline, and the waterline itself. I also submit the HDTV example I used at the end of Chapter 1. Whichever way one sees it, consciousness remains one of the greatest mysteries of life.

Consciousness has many uses, such as helping us concentrate on the many unexpected problems that crop up every day, problems that no automatic response can fully address. It also leaves plenty of room for automatic responses that can occur without any conscious effort or thought. For example, my fingers are moving across a keyboard in very deliberate movements designed to convey meaning through language. My movements are extremely precise, yet I am not aware of my brain telling my fingers to move. For that matter, I don't even consciously know how to make my body move, despite being adept at a wide array of motion requiring lots of coordination. As Karl Pearson said, the position of consciousness appears to be that of a helpless spectator of but a minute fraction of a huge amount of brain work. Voluntary movement that requires conscious intent, but operates without any conscious thought, is but one of many examples.

The highest form of ignorance is when you reject something you don't know anything about.
Wayne Dyer

Imagine having to think carefully about every move you make, lest you press the wrong button or lose your balance. I dare say our movements would be excruciatingly slow and difficult, to the point where the old cliche about walking while chewing gum would present an almost insurmountable challenge. Relegating these complicated-yet-routine functions to subconscious automatic routines saves us a lot of effort, and makes us faster, and better able to survive to boot. Consciousness, like any other trait, is only truly useful when the benefit of investing the effort outweighs the cost. This does not argue either for or against the possibility of a soul any more than a car with computers that handle running the engine and managing onboard systems argues for or against a driver (except that cars need drivers in order to perform any meaningful function).

Afterlife?

> *We have no reliable guarantee that the afterlife will be any less exasperating than this one, have we?*
> Noel Coward

To me, the question of whether there is an afterlife comes down to two fundamental topics: Eternal life, and the continuance of consciousness after physical death.

Eternal Life

Understanding the point I am about to make requires a brief explanation of Einstein's theory of relativity. Imagine that you are on a train traveling at a constant speed. As you gaze out the window, you feel no motion, but the world appears to be passing you by. A person standing near the tracks feels no motion, but the train appears to be passing them by. Who is correct? That all depends on whose point of view is used when framing the question. To the person on the train, the correct answer is "the world is passing me by" while the person on the ground's answer is "the train is passing me by."

> *Eternity is not something that begins after you're dead. It is going on all the time. We are in it now.*
> Charlotte Perkins Gilman

Assume for just a moment that you are simply a walking, talking, bag of meat. Your consciousness, your personal identity, did not exist before you were born, and will cease to be upon your death, meaning that there will no "you" to know that you ever even existed (a frightening thought, even to me). You may be aware of time existing before your birth (such as by studying history) and of time existing after you die (such as by learning of your children's future plans), but have not—and will not—experience these times for yourself. As far as your individual consciousness is concerned—from your individual point of view—time began the moment your brain "switched on" enough to realize its own existence, and will end the moment you die. From your own individual perspective, you will have therefore lived all the time there is. You will quite literally have lived an eternal life, because your time did not exist before you, and will not exist after you.

By this definition, eternal life does not require religion, superstition, a god, or a soul. All that's needed is relativity, and all that's needed to prove the validity of relativity is to imagine yourself and your reality from someone else's perspective. By this definition, each of us is living for all of our personal eternities. We all have eternal lives!

Death is the golden key that opens the palace of eternity.
John Milton

Let's throw the idea of a soul into this mix. If there is some continuation of consciousness following physical death—if our brains are indeed transceivers for consciousness and not the source—then we have no reason to assume that this life is our first, any more than we have to assume that it's our last. Do we reincarnate to keep on living over and over again until we reach enlightenment? Or does each life form some different chapter in an endless meta-life? Read *The Divine Savage* to get the answer.

Life after Death

Nothing about evolution or creation theory requires the presence of any soul or other continuance of consciousness following physical death. On the other hand, the world is awash in tales of near-death experiences, ghosts, sensing otherworldly presences, etc. Are they all bogus, fed by our desperate desire for something more and a veritable army of opportunistic charlatans, or is there something more to these tales?

The first requisite for immortality is death.
Stanislaw Lec

There seems to be no way to be sure without dying ourselves, an experience I am in no hurry to undergo—an attitude I assume you share. However, even if 99.999% of these accounts are bogus or otherwise explainable without any supernatural occurrences, well, that still leaves a lot of questions and possibilities open. Again, I don't know. I happen to believe in a form of life after death, but that's just me.

Chapter 25

Putting it all Together

Well, we've done it. We have boiled life down to its core functions of predator avoidance, group status, food, shelter, reproduction, and death. In so doing, we have exposed both the extreme simplicity of life, and the marvelously complex richness with which that simplicity manifests itself. Put another way, we have examined our daily lives in the context of evolution that continues to occur after billions of years. As we near the end of our journey, it's appropriate to look back at what we've learned to see just how far we've come in some ways, and how much we remain unchanged in others. We do this in order to answer the question I posed at the beginning of this book: Who are we, and what does it mean to be human? I can now answer that question in light of what I've presented in Chapters 1 through 24.

Animal Urges

All I know is that every time I go to Africa, I am shaken to my core.
Stephen Lewis

First and foremost, humans are animals. As animals, we are just as loyal to our genetically programmed biological urges as any other animal. Everything we do, from hanging out with friends to going to work, having sex, raising children, and dying, follows our fundamental programming to the core. The only real variable here is how exactly each individual goes about obeying her or his programming. And that's where our core beliefs and the process of realization described in Chapter 5 come into play. In other words, we are all following the same basic story line. One might be told as a Western, another as science fiction, and still another as comedy but the underlying plot is always the same. So what's the plot?

All human animals have the following goals in life:

- *Work*. Everyone wants to be productive, to know that she or he is contributing to the welfare of themselves, their families, and society in general. We work to provide food, shelter, and security for ourselves and those around us. We tend to reward those who pull their own weight, and look down on those perceived as lazy or ineffectual. This drive boils down to the biological imperative to reproduce and leave a genetic legacy that will endure long after we're gone.

- *Resources*. We seek to have our work yield enough resources on which to survive and even thrive. Most modern societies use money as the universal medium of exchange. It therefore follows that money is a powerful motivator, because it lets us secure the resources we need to eat and secure a home out of the elements. Money in particular also confers the additional privilege of status because we can use it to buy art, cars, homes, etc. that advertise our elite position to the rest of the world. Status in turn buys power, from which can flow leadership and/or enhanced opportunities to mate.

- *Health*. Taking care of ourselves is the best way to ensure that we will be able to both reproduce and take care of our offspring.

- *Love*. Who doesn't want to find that long-term pair bond that will give us the best possible chance of passing on our genetic legacy? Who does not seek the acceptance of parents and peers, in order to benefit from the safety and opportunities that only a group provides?

- *Fun*. One of the single largest benefits of our ever-advancing technologies has been an explosive growth in both our leisure time and ways to fill that time. From sitting around the campfire sharing tall tales of the hunt with fellow tribe members to catching the latest blockbuster movie, entertainment allows humans to forget

At the end of the day, the goals are simple: safety and security.
Jodi Rell

Just play. Have fun. Enjoy the game.
Michael Jordan

the daily grind for a time, and, most importantly, to build and maintain the social bonds that we depend on for our very lives.

- *Longevity.* Few if any people truly want to die. The majority of suicides aren't so much about wanting to stop living as about stopping the pain, getting even, or some other thing. Staying alive allows us to raise more children, and to contribute to raising our grandchildren—and right back to the holy grail of passing on the genes we go.

If this list sounds awfully similar to the list I presented in Chapter 1 of *The Enlightened Savage*, it is because people are people and understanding our universal goals and where those goals come from is a key step on the road to both achieving those goals and doing so in style.

The Substance of Style

Your success and happiness lies in you. Resolve to keep happy, and your joy and you shall form an invincible host against difficulties.
Helen Keller

How can you, a human animal, fulfill your animal urges in a way that offers joy and meaning to your life? I believe that my friend Caterina Rando nailed the answer with her fourteen success secrets, which I present here in a form adapted to the biological reality we've explored throughout this book.

Choose Joy

Joy descends gently upon us like the evening dew, and does not patter down like a hailstorm.
Jean Paul

Revel in the utter simplicity of life stripped of all its seeming complexity. You are here to pass along your genetic legacy (or at least to go through the motions). As we've seen throughout this book, you are a highly sexual, highly sensual, highly social creature. Enjoy it! Those who seek to deprive you or otherwise restrict your most fundamental instincts are only interesting in controlling you. Enjoy your friends and coworkers. Take pleasure from eating, sleeping, working, playing, or anything you do. Just never violate the fundamental social contract of "I won't harm you if you don't harm me," lest you suffer the consequences.

Live on Purpose

Life consists of six very simple core functions. How you carry out those functions is up to you. You may only have this one lifetime to live. It therefore behooves you and all around you to live with as strong a sense of mission as possible. Whether you're out to raise a family, save an endangered species, or build a corporation, do it with as much zeal, clarity of purpose, and sense of mission as you can muster.

> *The purpose of our lives is to be happy.*
> Dalai Lama

Acknowledge Others Often

Each of us depends on many other people for just about everything in life. This is part of fitting into the group, and obtaining or maintaining the status that is so critical to our lives as social animals. Treat others the way you want to be treated, and never fail to acknowledge the many services and kindnesses others give you.

Ask for What You Want

No person is an island, meaning that everyone needs help from time to time. If you are in need of something, ask for it. No pride, no long explanations, no negative self-talk, no obfuscation, nothing. Bare your soul, and ask for what you want. You will be amazed at how often you get what you seek. The catch here is that you must return favors, either to those who help you, or by helping others who ask you for help.

Be Willing to be Uncomfortable

There is nothing warm, fuzzy, cuddly, sweet, nice, or humane about life. Humans are the only animal that obsesses over how to kill the animals we eat—and each other—with as little pain as possible. Life is inherently uncomfortable. Those who insist on utter comfort are therefore not fully living. I'm not saying you need to thrust yourself into harm's way. I am saying that you need to be willing to take calculated risks from time to time. Going out on a limb can be both nerve wracking and very, very profitable.

> *As a child I was the best tree climber in our neighbourhood, I was like a little monkey. I've never been afraid of hurting myself or a little physical discomfort.*
> Rachel Weisz

Explore New Possibilities

> *Dream on it. Let your mind take you to places you would like to go, and then think about it and plan it and celebrate the possibilities. And don't listen to anyone who doesn't know how to dream.*
> Liza Minnelli

Humans are some of the most creative animals on this planet, if we'll just let ourselves use that creativity. I will even go so far as to say that creativity is what enabled us to evolve as far as we have. People migrating to new areas found different foods and encountered different dangers. How did we surmount those challenges? Innovation. We could not possibly have spread across the Earth without constantly exploring new possibilities.

Maintain a Positive Disposition

Always look at the bright side of life. You may as well, because this life is short. Challenges can pose either risks or opportunity, depending on you see them. Remember that our highly developed frontal lobes create a version of reality that has been filtered through our beliefs. Strive to build and maintain positive beliefs, and your entire life will literally change. *The Enlightened Savage* talks about this in great detail.

Take Small Actions

The longest treks begin with a single step. The tallest buildings begin by turning over a single shovelful of dirt. Focus on every small step, and celebrate accomplishing that step as it happens. You'll achieve great things almost without realizing it.

Openly Express Your Gratitude

> *Develop an attitude of gratitude, and give thanks for everything that happens to you, knowing that every step forward is a step toward achieving something bigger and better than your current situation.*
> Brian Tracy

Whether it's the person who gives you your first big break, or the clerk who rings up your groceries, your life is better because of these people. Accept the gifts that are coming to you without preconditions or judgment, and be thankful. "Thank you" are the two most powerfully uplifting words in any language, provided they are said simply and sincerely. Being openly grateful rewards the person who helped you both my making them feel good, and by also enhancing the esteem with which others view your helper. This gives status to that person, which can go a long way toward improved standing within the group. Even better, your gratitude signals that you are a group player, and that you recognize the value of others. That can only help your own standing within the group.

Have Some Fun

All work and no play will make you one dull, stressed-out individual. Work your fingers to the bone, and you'll have bony fingers. You get the idea. Figure out the most efficient way to do whatever it is you need to do, and spend the rest of the time enjoying the fruits of your labors. Even our primordial ancestors recognized the benefits of down time.

Smile, Laugh, and Love More

Whatever you put energy into grows! If you don't have all the happiness, laughter, and love you want in your life, put that energy out there. Smiles and laughter are appeasement gestures that melt away aggression and fear like a torch melts ice. Love in all its forms strengthens family, group, and romantic bonds. People are drawn to—and want to be with—and help happy people.

A gentle word, a kind look, a good-natured smile can work wonders and accomplish miracles.
William Hazlitt

Look for the Ease

Never forget that work is a measure of results, not effort. Keep this in mind at all times. Got a task to do? Starting a project? What's the easiest way to get it done? If you look for ways to get results smarter instead of harder, you'll find that success really does come easy. Since you haven't expended all of your energy getting some success, you'll have plenty left that you can use to go out and get even more. Imagine getting more for less. You do that while shopping, don't you?

Try to be like the turtle—at ease in your own shell.
Bill Copeland

Always Expect Success

The process of realization described in Chapter 5 means that you will experience whatever reality you believe you will experience. You may as well believe in your own success, for it's both easier and more productive than believing in failure (which does not exist except as any discrepancy between what you think you want and what you really want/strive for).

You Have the Power

You create your own personal reality. You are responsible for your own life. You can change your beliefs. You can achieve great success with great ease. You can fulfill your life's mission

Dream no small dreams for they have no power to move the hearts of men.
Goethe

The Meaning of Humanity

and secure lasting freedom for yourself and your entire family for many future generations. No one else can do this but you.

> *Is the system going to flatten you out and deny you your humanity, or are you going to be able to make use of the system to the attainment of human purposes?*
> Joseph Campbell

What does it mean to be human? On one level, being human means being a simple prey animal that got exceptionally lucky (from an evolutionary perspective) and expanded across the globe, developing technology, that provides the illusion that we are both a third kingdom of life above plants and animals, and above the workings of evolution. I realize that this statement cuts my species down several sizes and knocks more than a few of us off our self-made pedestals. So be it. I believe that acknowledging our animal natures and fashioning our societies around those urges while eschewing attempts to restrict them in order to wield control, can go a long way to solving the challenges we face as a species. Other have said and I'll say it myself: If we fail as a species, it will be because of our extraordinary success, coupled with forgetting our very humble animal roots.

On another level, being human means having the trappings of incredible complexity while living out just six core functions, five of which exist either directly or indirectly to serve the prime function of reproduction. Your job, house, car, possession, and all of society are little more than illusions designed to carry out those six core functions.

> *In recognizing the humanity of our fellow beings, we pay ourselves the highest tribute.*
> Thurgood Marshall

On a third level, humans are the masters of our own lives to an unprecedented degree. We are the first animal capable of altering entire ecosystems, and of ending all life on this planet. Our unique brain structure means that we can live whatever kind of life we believe we can live, with the laws of physics being our only limitations (and even those may be pliable, according to at least a few people). Yes, we are humble prey animals. Yes, our lives consist of six core functions. Still, we humans alone control how we carry out those functions.

What does it mean to be human? Combine all of the three levels I just described, and the answer becomes: Being human means whatever each of us wants it to mean. I believe that following the strategies outlined above can help us find and build enduring personal meaning in each of our lives.

320 | The Natural Savage
Discovering the Human Animal

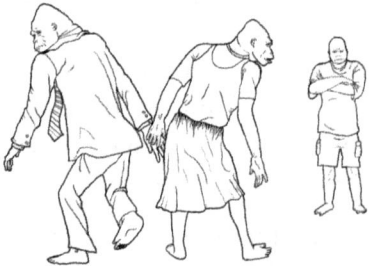

Chapter 26

Who Knows?

> *What's an expert? I read somewhere, that the more a man knows, the more he knows, he doesn't know. So I suppose one definition of an expert would be someone who doesn't admit out loud that he knows enough about a subject to know he doesn't really know how much.*
> Malcolm Forbes

Dear Reader,

I want to believe that this book provides a model for answering that question, both thoroughly and with adequate thoroughness for most readers. That said, wanting something to be a certain way does not necessarily make it so. Writing *The Natural Savage* has plunged me headlong into a debate that has raged for thousands of years, one that shows no sign of winding down any time soon. I am confident that people will debate my answers for solid scientific, religious, and philosophical reasons. and am always willing to amend—and even upend—model based on new and compelling information.

That does not mean that my writing this book or your reading it has been in vain. Far from it. Having a working model helps explain and provide much-needed context for observed behavior. Understanding the six core functions of life (predator avoidance, group status, food, shelter, reproduction, and death) helps expose the true motives behind people's words and actions. Memorizing these functions, plus knowing how they relate to each other, allows one to make sense of life by cutting through all of the clutter to arrive at what's going on below the surface. The potential benefits are enormous.

Knowing that humans have a mixed reproductive strategy may induce you and your significant other to come to an agreement that preserves your overall relationship while addressing the urge to wander. Knowing that reproduction trumps all other functions can help you spot people who are trying to control that fundamental drive in order to exert undue influence. If you know that anger is a response to fear, then you may be able to see your boss yelling at you in a different light. Having an understanding of the power of groups and the primal need for acceptance helps one understand why teenagers dress so oddly, listen to noise that you would never dare call music, and generally act like they own the world. Accepting that humans are prey animals with highly developed forebrains that create schemas out of raw sensory input may help further our understanding of mental health.

Perhaps most importantly, deflating our egos enough to admit that we are apes, glorious apes perhaps, but apes nonetheless, may give our species a much-needed dose of humility. This humility might prompt us to reexamine our relationship with the only planet in the universe known to harbor life, and to do a better job of preserving it for the all-important future generations. It might also free us from the tyranny of people who claim to have some special connections with a deity that created us right down to our biological urges, only to warn us against using those instincts, lest some eternal damnation befall us (unless we fight in wars, of course). If God or some other creator built us from the ground up (literally) and imbued us with these instincts, then I for one believe that the greatest praise we can bestow on our maker is to be true to the instincts "he" created. Refusing to do so is a bit like condemning a car to the scrap heap unless it operates like a boat.

And finally, accepting this model of who and what we are will help us realize both the immutable nature of the six core functions and the infinite variety of ways in which we can express those functions, thereby creating the kinds of meaningful lives we all want and richly deserve.

Everything you've read in this book represents my best attempt at beginning to answer one of the greatest questions of all time: What is the nature of the human animal? Having come this far, you are entitled to know my answer to this question and here it is:

I don't know. But then again, neither does anybody else.

Every book is a new journey. I never felt I was an expert on a subject as I embarked on a project.
David McCullough

Resources

I drew boundless inspiration for this book while reading the books and visiting the Web sites listed below. While this list comprises my main body of research, it is not intended as a complete bibliography by any means. The fact that you've read *The Natural Savage* this far can only mean that you, like me, find the subject of human nature fascinating. I therefore encourage you to both read the books and visit the Web sites for yourself, and avail yourself of the latest thinking and research into the human animal and the answer to the eternal question: Who are we, really?

Books

- *And The Animals Will Teach You*—Margot Lasher
- *Animals In Translation*—Temple Grandin & Catherine Johnson
- *Bonobo: The Forgotten Ape*—Frans De Waal & Frans Lanting
- *Evolutionary Thought in Psychology*—Henry Plotkin
- *In the Shadow of Man*—Jane Goodall
- *Kant and the Platypus*—Umberto Eco
- *Kanzi: The Ape at the Brink of the Human Mind*—Susan Savage-Rumbaugh & Roger Lewin
- *Manwatching*—Desmond Morris
- *Molecules of Emotion*—Candace Pert
- *Muddling Through: Pursuing Science and Truth in the 21st Century*—Mike Fortun & Herbert Bernstein
- *On Human Nature*—Edward O. Wilson
- *Our Inner Ape*—Frans De Waal
- *Patterns of Culture*—Ruth Benedict
- *Peacemaking Among Primates*—Frans De Waal
- *Psychology: An Evolutionary Approach*—Steven J. C. Gaulin & Donald H. McBurney

- *Science and Evidence for Design in the Universe*—Michael J. Behe, et. al
- *Sex, Power, Conflict: Feminist and Evolutionary Perspectives*—edited by D.M. Buss and N. Malamuth
- *Sexual Behavior in the Human Male*—Alfred Kinsey, et. al
- *Sexual Behavior in the Human Female*—Alfred Kinsey, et. al
- *Sociobiology*—Edward O. Wilson
- *The Ascent of Mind*—William H. Calvin
- *The Complete Idiot's Guide to Psychology*—Joni Johnston
- *The Fermi Solution*—Hans Christian von Bayer
- *The Game*—Neil Strauss
- *The God Delusion*—Richard Dawkins
- *The Human Zoo*—Desmond Morris
- *The Mother Tongue*—Bill Bryson
- *The Naked Ape*—Desmond Morris
- *The Third Chimpanzee*—Jared Diamond
- *The Wonder of Being Human: Our Brain & Our Mind*—Sir John Eccles & Daniel N. Robinson
- *Time, Love, Memory*—Jonathan Weiner
- *UnHypnosis*—Steve Taubman
- *Unwritten Rules of Social Relationships*—Temple Grandin & Sean Barron
- *Walking With the Great Apes*—Sy Montgomery
- *When Elephants Weep*—Jeffrey Masson & Susan McCarthy
- *Whis is Sex Fun?*—Jared Diamond
- *Why We Love*—Helen Fisher

Web Sites

- *Kinsey Institute*—www.indiana.edu/~kinsey
- *Wikipedia*—www.wikipedia.com (original sources cited where appropriate)
- *Online Dictionary*—www.dictionary.com (original sources cited where appropriate)

The Natural Savage
Discovering the Human Animal

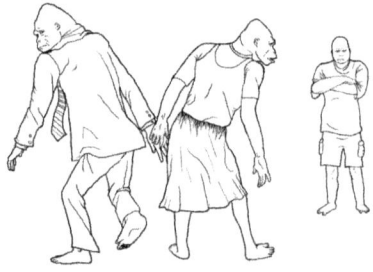

www.ingramcontent.com/pod-product-compliance
Lightning Source LLC
Chambersburg PA
CBHW080238170426
43192CB00014BA/2482